T0074324

Corrosion of
High-Performance Ceramics

Yu. G. Gogotsi · V. A. Lavrenko

Corrosion of High-Performance Ceramics

With a Foreword by F. Thümmler

With 124 Figures

Springer-Verlag
Berlin Heidelberg New York
London Paris Tokyo
Hong Kong Barcelona
Budapest

Dr. Yury G. Gogotsi

Institut für Keramik im Maschinenbau, Universität Karlsruhe, Postfach 69 80,
W-7500 Karlsruhe, Fed. Rep. of Germany

Professor Dr. Vladimir A. Lavrenko

The Institute for Problems of Materials Science, Ukrainian Academy of Sciences,
252680 Kiev, Ukraine

Translator:

T. A. Maximova

Ukrainian Pulp and Paper Research Institute, 18/7, Kutuzov St.,
252133 Kiev, Ukraine

ISBN-13:978-3-642-77392-1

Library of Congress Cataloging-in-Publication Data. Gogotsi, IU. G., 1961 - Corrosion of high-performance ceramics / Yu. Gogotsi, V. A. Lavrenko. p. cm. Includes bibliographical references and index. ISBN-13:978-3-642-77392-1   e-ISBN-13:978-3-642-77390-7
DOI: 10.1007/978-3-642-77390-7   1. Ceramic materials--Corrosion. I. Lavrenko, V. A. (Vladimir A.), 1933- . II. Title TA455.C43G64   1992   620.1'404223--dc20   92-13133

© Springer-Verlag Berlin Heidelberg 1992
Softcover reprint of the hardcover 1st edition 1992

Production Editor: A. Kübler
Typesetting: Data conversion by Springer-Verlag
54/3140 - 5 4 3 2 1 0 - Printed on acid-free paper

Foreword

Engineering ceramics are often subjected to high thermal loads over long periods of time, which mainly lead to multiaxial stresses. This may result in a degradation of the material properties with time and temperature. The resulting engineering problems of lifetime and reliability are often difficult to solve. Apart from commercial considerations, this is the main reason why the introduction of engineering ceramics in technology is advancing more slowly than formerly anticipated. In addition to the above-mentioned problems, nonoxide ceramics are also subject to oxidation or corrosive attack, which is of enormous importance under long-term, high-temperature service conditions. They are thermodynamically unstable in air and their survival depends largely on the existence of protective oxide layers. Oxidation and corrosion processes strongly depend on internal and external parameters and lead to an increased danger of degradation of the mechanical properties.

The authors *Yu. G. Gogotsi* and *V.A. Lavrenko* are outstanding experts in the field of corrosion of engineering ceramics. They have been working in this field for a long time and thus have gathered wide experience. For several years their book "Corrosion of Structural Ceramics", published in Russian, was the only monograph in this field. The decision of Springer-Verlag to provide an English edition created the opportunity and the necessity for the authors to prepare a new, updated manuscript which is now presented to the scientific and engineering community. This monograph provides a comprehensive overview, emphasizes the relevance of this field for many present and future applications, and may help to develop ceramics of improved oxidation and corrosion resistance.

Prof. Dr.-Ing. habil. F. Thümmler
Institut für Werkstoffkunde II
Universität Karlsruhe

Preface

One of the most important problems of materials science is the development of advanced materials that operate reliably at high temperatures, under mechanical loads and in aggressive environments. Under such extreme conditions ceramic materials are more efficient than metals. Ceramics based on carbides, nitrides and oxides possess a unique combination of properties: average density, low thermal linear expansion coefficient, high corrosion and erosion resistance, hardness and satisfactory strength above $1000°$ C. The reserves of raw materials suitable for manufacturing ceramics are virtually unlimited and their costs under large-scale production conditions can be lower than the costs of nickel, cobalt, tungsten, titanium and other components of oxidation-, and corrosion-resistant superalloys.

The performance of structural ceramics is mainly determined by their mechanical properties and corrosion resistance. However, while quite a number of monographs, reviews and articles are devoted to the mechanical behaviour of structural ceramics, the problems of their corrosion have never been summarized in a sufficiently comprehensive form. Some data on the corrosion of oxide and nonoxide structural ceramics in liquid and gaseous environments can be found in handbooks and monographs describing several types of materials based on refractory compounds, but the relation between corrosion resistance and mechanical behaviour has not been examined in these publications. Our monograph "Corrosion of Structural Ceramics" (Metallurgiya, Moscow 1989) was the first attempt to summarize both the literature and the authors' own results of investigations of corrosion and its effect on the properties of structural ceramics.

To be more concise and to deal with more urgent problems, the present edition does not contain the chapters devoted to the processing of ceramics, experimental methods of investigation, mechanisms of gaseous and molten metal corrosion and information on the oxidation of nitride and carbide powders. However, the book now provides data on ZrO_2 ceramics and ceramic matrix composites. Moreover, all the chapters were updated to do justice to recent publications. Because of the limited size, only the most promising silicon nitride and silicon carbide materials are discussed in detail. We pay less attention to aluminium nitride, boron nitride and boron carbide materials, which possess various valuable properties but do not find many applications as structural materials. We examine only some aspects of oxide ceramics which display higher corrosion resistance and find wide commercial application, because their properties are described thoroughly enough in the literature. We have updated and supplemented the list of references, but nevertheless it is not complete and only selective.

At present the properties of structural ceramics are being investigated in Germany, the USA, Japan, the UK, France, Italy, the former USSR and other countries, the results being published in tens of papers every year. Many aspects of the problem have not been studied thoroughly enough and we would like to draw the attention of researchers to unsolved problems. This book will be of use for process engineers, chemists and other specialists who work on the development of structural ceramics and study their properties and commercial applications.

All comments and suggestions on the content of the book will be most welcome by the authors and taken into consideration in their future work.

Karlsruhe, Kiev, May 1992   Yu.G. Gogotsi, V.A. Lavrenko

Acknowledgements

The authors would like to thank the research staff of the Kiev Polytechnical Institute, the Institute for Problems of Materials Science and the Institute for Problems of Strength of the Ukrainian Academy of Sciences, as well as other colleagues who took part in the investigations which became the basis of this monograph. We are particularly grateful to Dr.Sci. G.A. Gogotsi and Dr. V.P. Zavada of the Institute for Problems of Strength, Dr. V.P. Yaroshenko of the Institute for Problems of Materials Science (now at NEOX Co.) and Dr. F. Porz of the Institut für Keramik im Maschinenbau, University of Karlsruhe, who have kindly acted as "scrutineers" by reading one or more chapters. All of them pointed out numerous errors in the drafts and made valuable suggestions, most of which were adopted or utilized indirectly. Yu.G. Gogotsi would like to take the opportunity to acknowledge his gratitude to Prof. F. Thümmler and Dr. G. Grathwohl for their support during his work at the University of Karlsruhe. Yu.G. Gogotsi is also thankful to the Alexander von Humboldt Foundation for financing his stay in Germany where this monograph was being prepared for publication.

Contents

List of Abbreviations

AES	–	Auger electron spectroscopy
CVD	–	chemical-vapour deposition
DTA	–	differential thermal analysis
DTG	–	differential thermogravimetry
EDAX	–	energy-dispersive X-ray microanalysis
HIP	–	hot isostatic pressing
HIPSN	–	hot isostatic pressed silicon nitride
HP	–	hot pressing
HPSiC	–	hot-pressed silicon carbide
HPSN	–	hot-pressed silicon nitride
IR	–	infrared
MOR	–	modulus of rupture (three-point bending strength)
PSZ	–	partially-stabilized zirconia
RBSN	–	reaction-bonded silicon nitride
SCC	–	stress corrosion cracking
SEM	–	scanning electron microscopy
SIMS	–	secondary ion mass spectrometry
SRBSN	–	post-sintered reaction-bonded silicon nitride
SSN	–	sintered silicon nitride
TG	–	thermogravimetry
TZP	–	tetragonal zirconia polycrystals
XRD	–	X-ray diffraction
XPS	–	X-ray photoelectron spectroscopy
ZTA	–	zirconia-toughened alumina

1. Introduction

The advance of science and technology is inseparably connected with improving the existing structural materials and developing new ones. In mechanical engineering, chemical industry, metallurgy and other industries higher oxidation resistance, high-temperature strength and corrosion resistance become more and more essential properties for materials operating in different aggressive environments at high temperatures.

Since the potentials of metal alloys have practically reached their limit, special importance is attached to ceramics as a promising material for manufacturing structural components operating at temperatures up to 1400° C and higher. Ceramic materials can retain their high strength above 1000° C [1.1, 2] and are capable of operating in aggressive and oxidizing environments [1.3, 4].

At present ceramic materials are widely applied in many industries. They are used for manufacturing heat exchangers, components of paper machines, chemical and metallurgical equipment [1.5, 6]. But their advent as substitutes for metals in engines is considered the most advantageous [1.7, 8]. A gas-turbine engine with ceramic components can operate at higher temperatures and with higher efficiency. Ceramics used in a diesel engine [1.9] make it possible to run it uncooled, they can reduce mass and improve efficiency.

The application of ceramics in engines is determined not only by their high oxidation resistance but also by their higher corrosion resistance compared with metals, which make it possible to use low-grade fuels.

Ceramics are used for manufacturing stators, rotor blades, flame tubes of combustion chambers and other components of gas-turbine engines as well as pistons, cylinders, fuel valves, swirl chambers and other components of diesel engines [1.8, 10].

At present nonmetal carbide- and nitride-based ceramics are considered the most promising for structural applications [1.11–14]. These are compounds mostly formed by covalent bonding, their crystals are characterized by considerable Peierls stresses determined by the resistance to bond distortions inherent in their lattice. Therefore they exhibit high stability in terms of dislocation motions and retain their strength up to very high temperatures. Although nonoxide structural ceramics have several advantages compared with oxide ceramics they possess one essential limitation, viz. they oxidize on heating in air up to high temperatures. The strength of oxide ceramics which are resistant to oxidation, degrades at lower temperatures more severely than that of nonoxide ones [1.15, 16]. Their reinforcement with particles, fibers or whiskers of refractory compounds makes them also sensitive to oxidation. If we take into account that under service conditions the engine parts are not only oxidized by hot air but also affected by corrosive combustion products and sea salts injected by marine-based turbine engines, it becomes quite understandable that these ceramics should be resistant to high-temperature corrosion. Thus, the corrosion resistance is one of the most important factors determining the applicability of ceramics for manufacturing engine components.

Corrosion of ceramic materials is understood (as in the case of metals) as a damage of structural components due to chemical reactions of the ceramics with the substances present in the environment, i.e. one should consider the reactions between ceramic materials and melts, solutions or gases occurring at the interface.

It is necessary to note that the rather high corrosion resistance of ceramics makes it very difficult to estimate the extent of corrosion damage by mass variation of specimens, the depth of attack, the number of corrosion zones per unit surface area, etc., as it can be done for metals. Moreover, oxidation can result not only in the deterioration of properties but also in a higher strength of ceramics and, thus, in a higher performance of a ceramic product. Therefore the effect of corrosion on the performance of structural ceramics can be estimated only by a variation of their mechanical properties. The complex processes at high-temperature corrosion and various influencing factors determine the need for a comprehensive approach to this problem.

While corrosion of engineering materials (metals and alloys) has been investigated in a great number of studies, and several fundamental monographs were devoted to this problem [1.17–19], high-temperature corrosion of ceramics was virtually studied only in the past 15–20 years. The investigations of silicon carbide materials designed to be applied as refractories or high-temperature heaters and oxide refractory materials were a certain exception.

Moreover, the Symposium on Corrosion of Ceramics and Advanced Materials held within the Electrochemical Society Meeting in Washington in May 1991 [1.20] and several other conferences devoted to this problem [1.21, 22] indicate that at present studies of the oxidation and corrosion of ceramics are being established as an independent scientific trend.

The present monograph summarizes the results of recent investigations of corrosion and its effect on the properties of ceramics based on silicon nitride, silicon carbide, boron carbide, aluminium nitride, boron nitride, zirconia, alumina and mullite. The behaviour of these ceramics is always analysed taking into account their real properties, composition and structure. The structure and properties of ceramic materials are determined to a considerable extent by the production technology. In this book we cannot dwell on the production technology of ceramics; however, the data will always be presented with respect to processing and in many cases we examine separately the oxidation of ceramics produced by a certain method, e.g. hot pressing or reaction bonding. The processing of the examined materials is described in more detail elsewhere [1.7, 12–14, 23–25].

2. Oxidation in Oxygen and Air

High-temperature oxidation is the most wide-spread type of corrosion of structural ceramics. It is similar to the oxidation of metals in many respects. The same methods and theories are used to investigate and describe the kinetics of both processes. At the same time there are also several differences. Thus, due to the very high oxidation resistance, the majority of ceramic materials should be investigated with the help of precision instruments; the dielectric properties of ceramics restrict the applicability of certain modern physico-chemical methods of analysing oxidation products; the oxidation of nonoxide ceramics results both in solid and gaseous products, which causes certain difficulties in quantitative calculations following the theories developed for metals.

The problem of investigating the high-temperature oxidation of ceramics was posed by practical requirements. The necessity to evaluate the most important properties of ceramics developed for engines and other devices prompted researchers to perform special investigations of their oxidation. Therefore, the majority of data on the oxidation resistance of structural ceramics was obtained simultaneously with the investigations of strength, creep and other properties of ceramics. Specimens of very different size and shape and various techniques were used for this purpose [2.1 et al.], while even the material of the crucible [2.2] and the type of furnace [2.3] can influence the results. Therefore, such data are difficult to compare and are only of informative value.

On the other hand, in some investigations of the general mechanisms of oxidation the notions "chemical compound" and "ceramics" were erroneously identified. Whether powders [2.4] or compact ceramics specimens [2.5] were investigated, they were called "silicon nitride", "silicon carbide" etc., regardless of the fact that the densification of refractories changes their properties entirely due to the addition of sintering aids, certain variations of phase composition, grain sizes, the surface state and so forth. The presence of tenths of a percent of sintering aids or impurities in ceramics can already lead to crucial changes in their oxidation behaviour. Thus, without taking into account the salient features of ceramic processing and the other factors mentioned above, one cannot judge the actual mechanisms of oxidation processes. Therefore, the data on the oxidation resistance of ceramics cited in [2.6], which were obtained without taking into account the peculiar features of the studied specimens are of limited practical value. In [2.7, 8] the data on the processing of the materials are summarized, but they are rather limited.

In the following, if we do not mean chemical compounds in general but ceramics, their production technology will always be specified (e.g. hot-pressed

B_4C, reaction-bonded Si_3N_4). Expressions like "Si_3N_4-based ceramics" or "silicon nitride ceramics" will always refer to the whole class of silicon nitride-based materials, etc.

2.1 Silicon Nitride Ceramics

Silicon nitride ceramics, which are mainly used for engine components, are the focus of our attention.

2.1.1 Thermodynamics of Si_3N_4 Oxidation

Thermodynamic calculations of equilibria in the system $Si_3N_4 - O_2$ performed by *Vojtovich* [2.9] at a pressure of 0.105 and 1.3 Pa demonstrated that a large number of reactions resulting in the formation of solid (SiO_2, Si_2N_2O, SiO) and gaseous (N_2, N_2O, NO, NO_2, SiO, SiN) products are possible. The reactions leading to the formation of silica, nitrogen and/or its oxides are characterized by the most negative change of Gibbs energy, but their thermodynamic probability decreases with temperature. At high temperatures the possibility of reactions leading to the formation of SiO increases. However, up to 2000° C and at atmospheric pressure the following reaction is the most probable:

$$Si_3N_4 + 3\,O_2 = 3\,SiO_2 + 2\,N_2 \quad . \tag{2.1}$$

In order to determinate, which predominates under real conditions, the gaseous oxidation products were investigated by IR spectroscopy [2.10] at a pressure of 0.1 Pa and by mass-spectrometry [2.11] at an oxygen pressure of 0.023 Pa. The authors [2.10] found nitrogen dioxide and stated the reaction

$$Si_3N_4 + 7\,O_2 = 3\,SiO_2 + 4\,NO_2 \quad . \tag{2.2}$$

Mass-spectroscopic investigations revealed O_2^+, N_2^+, NO^+, N_2^{2+}, O_2^{2+} and made it possible to conclude [2.11] that the reaction

$$Si_3N_4 + 5\,O_2 = 3\,SiO_2 + 4\,NO \tag{2.3}$$

is possible. During oxidation of sintered silicon nitride (SSN) the authors [2.12, 13] also found N_2 and NO.

The question, which of the three reactions (2.1–3) prevails is very complicated and has not yet been solved, though reaction (2.1) is generally used to describe Si_3N_4 oxidation. The formation of silica was reported by most of the researchers [2.14, 15]. There are data suggesting that at temperatures up to 900° C in air the surface of a specimen is covered with solid silicon monoxide formed as a brown phase with the inclusions of bright-red crystals [2.9]. However, this was not confirmed by other investigations. It is reported in [2.16] that at temperatures above 1500° C the reaction

$$Si_3N_4 + 3\,SiO_2 = 6\,SiO_{(gas)} + 2\,N_2 \tag{2.4}$$

takes place. This assumption is based on the fact that during high-temperature oxidation the surface of specimens becomes not smooth but blistered due to vigorous gas release through the oxide film. However, at temperatures above 1420° C Si₃N₄ starts to dissociate into elements, here silicon reacts with SiO₂ forming gaseous silicon monoxide. According to the data provided in [2.17] sintering aids added to the materials catalyse the processes occurring on their oxidation. Thus, when 5% NaF is added, silicon oxynitride is formed instead of SiO₂. The presence of magnesium in the material promotes the oxidation of Si₃N₄ to SiO₂ [2.18]. At the same time it is reported in [2.19] that both MgO and Y₂O₃ facilitate the formation of Si₂N₂O from Si₃N₄ and SiO₂.

Thus, process conditions and additives in ceramics determine the formation of different oxidation products.

For describing the oxidation of silicon nitride materials the Wagner theory is applicable [2.15]. However, the description of the process is complicated by the fact that both solid and gaseous oxidation products are formed. The application of the Wagner theory with certain assumptions made it possible to calculate that the oxidation resulting in silicon monoxide formation in accordance with the reaction

$$2 \, Si_3N_4 + 3 \, O_2 = 6 \, SiO_{(gas)} + 4 \, N_2$$

occurs, e.g. at 1300° C only when the oxygen pressure is below 80 Pa [2.15]. The plot showing the areas of "active" Si₃N₄ oxidation when only gaseous substances are formed and those of "passive" oxidation when solid products are formed is presented in Fig. 2.1. As regards solid products, there are different opinions as to the limiting factor of the oxidation process: the diffusion of oxygen to the Si₃N₄/SiO₂ interface [2.21], of nitrogen to the SiO₂/air interface [2.22] or of additives and impurity ions from inner layers to the surface of ceramics [2.23].

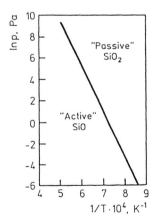

Fig. 2.1. Calculations of partial oxygen pressures on the transition from "active" to "passive" silicon nitride oxidation [2.20]

2.1.2 Composition of Oxide Layer

In the majority of cases the oxidation of Si_3N_4 ceramics results in a silica layer on the surface of the specimens.

Investigations of Si_3N_4 powders have shown that up to 1065° C amorphous SiO_2 is formed [2.2, 4]. Formation of α-cristobalite was found at higher temperatures by many researchers [2.14, 25]. Cristobalite was also reported for the oxidation of silicon oxynitride. The studies on the oxidation of Si_3N_4 powders and ceramics also revealed tridymite and quartz [2.24] above 1125° C and 1385° C, respectively. In the literature there are no data on the formation of other SiO_2 modifications. In the surface layer of oxidized specimens silicon oxynitride [2.26] as well as impurities and additives [2.27] with a high affinity for oxygen can also be present, with their contents well exceeding their average concentrations in ceramics. Thus, according to [2.15], after the oxidation of the material with a composition of $Si_3N_4 + 1\%\,MgO$ at 1400° C in dry oxygen the protective film consisted mainly of enstatite $MgSiO_3$. The authors of [2.28] discovered up to 40% calcium in an oxidized surface layer, while the calcium content in the bulk material was below 1%. In the oxide film on yttria-doped specimens $Y_2Si_2O_7$ can usually be found and in that on ceria-doped specimens one finds CeO_2 [2.29]. The migration of calcium, magnesium and other impurities to the surface of a specimen results in silicate phases, their softening temperature being between 1100° and 1300° C [2.18]. The dissolution of Si_3N_4 in this liquid phase increases its viscosity and promotes the formation of a vitreous film on the surface of a specimen [2.18]. In contrast to this, the presence of rare-earth oxides or water vapours in a reaction atmosphere facilitates the crystallization of an oxide layer [2.30].

Thus, the composition of the oxide layer is determined by the oxidation conditions and the composition of the ceramics. Here it is necessary to take into account that the phase transitions in silica depend to a large extent on temperature, cooling rate and the presence of impurities. It is precisely these factors that apparently determine if the silica formed on oxidation is present in an amorphous state or in a crystalline form. However, the existing data are rather contradictory [2.24, 31] and do not allow to give an unambiguous answer to the question when the one or other SiO_2 modification is formed. On multi-component materials there will always emerge an oxide layer containing different phases (Fig. 2.2).

2.1.3 Effect of Phase and Chemical Composition of Ceramics

Silicon nitride exists in two polymorphic modifications similar in structure and differing only in the order of arrangement of tetrahedral elements in the direction of the crystallographic axis c [2.33]. The nitridation of silicon at 1200°–1400° C results predominantly in α-Si_3N_4, and above 1450° C β-Si_3N_4 is formed. Between 1450° and 1500° C a slow, irreversible $\alpha - \beta$ transformation takes place [2.33]. Nearly all the commercial silicon nitride powders are a mixture of α- and β-modifications.

6

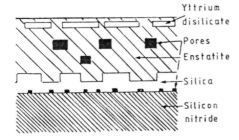

Yttrium
disilicate
Pores
Enstatite
Silica
Silicon
nitride

Fig. 2.2. Schematic diagram of phase distribution in an oxide film on MgO- and Y_2O_3-doped silicon nitride [2.32]

Studies on the oxidation of specimens with different contents of α- and β-Si_3N_4 [2.9] have demonstrated that when all other conditions are the same, the specimens which consist mainly of α-Si_3N_4 are more susceptible to oxidation. With an increase in oxidation temperature the content of α-Si_3N_4 in tested specimens decreased. In the opinion of *Engel* [2.34], this is due to the predominant oxidation of an α-phase. At the same time a similar effect may be associated with the $\alpha - \beta$ transformation on heating.

During chemical-vapour deposition, besides α- and β-Si_3N_4, amorphous silicon nitride is formed, the oxidation resistance of which is higher than that of the crystalline modifications [2.35].

As mentioned above, the chemical composition also has considerable influence on the oxidation process. Almost every silicon nitride ceramic contains one of the sintering aids: MgO [2.36], Al_2O_3 [2.37], Y_2O_3 [2.29], BeO [2.38], CeO_2 [2.29], ZrO_2 [2.39], Yb_2O_3 [2.8] etc., but magnesia, alumina and yttria are the most widely used ones. After oxidation of MgO-doped ceramics, their surface layer reveals enstatite and forsterite Mg_2SiO_4 [2.22] formed as a result of the reactions:

$$2\,MgO + SiO_2 = Mg_2SiO_4 \quad , \tag{2.5}$$

$$Mg_2SiO_4 + SiO_2 = 2\,MgSiO_3 \quad . \tag{2.6}$$

More detailed information on the effect of sintering aids on the oxidation resistance of silicon nitride ceramics will be given later.

2.1.4 Oxidation of Porous Ceramics

When reaction-bonded Si_3N_4 (RBSN) with a porosity of usually 15%–30% [2.40] is oxidized, the process takes place in the bulk of a specimen, with considerably higher oxidation rate than in the case of dense ceramics. Thus, according to the data provided in [2.41], after oxidation of RBSN with a density of $2270\,kg/m^3$ at $1040°$ C for $200\,h$ the specimen reveals 28% SiO_2, while its content in the initial material did not exceed 4%. According to [2.42], the dependence of the oxidation rate on porosity is linear and can be expressed by the equation

$$\Delta m = kS + kV f_p (1 - \gamma_o)/\bar{R} \quad , \tag{2.7}$$

7

where Δm is the mass gain; S is the geometric surface area of a specimen; f_p is the fraction of open pores (of total porosity); $\bar{R}$ is the average pore radius; γ_o is the relative density; V is the volume of a specimen; k is the coefficient.

It follows from the equation that not only the total open porosity but also the pore radii are of importance. However, it is quite possible that it is not the pore radii but in fact the radius of the channels connecting the pores that exerts the major influence on the oxidation resistance. The filling of these channels with an oxide phase with a larger volume than the oxidized compounds (Table 2.1) leads to a retardation of the inner oxidation of ceramics. It should be noted that the oxidation of a porous material results in a higher density and smaller porosity.

Table 2.1. Mass and volume changes on complete oxidation [2.21]

Crystallochemical transformation	$\Delta m, \%$	$\Delta V, \%$
$Si \rightarrow SiO_2$	113.9	116.7
$SiO \rightarrow SiO_2$	36.3	26.3
$SiC \rightarrow SiO_2$	49.9	107.4
$Si_3N_4 \rightarrow SiO_2$	28.5	76.1
$Si_2N_2O \rightarrow SiO_2$	19.9	44.2

Though the Si_3N_4 oxidation process obeys the Wagner theory, the kinetic curves of the oxidation of porous RBSN as opposed to dense hot-pressed silicon nitride (HPSN) or SSN [2.9] cannot be described by the parabolic equation

$$(\Delta m/S)^2 = K_p t + A \quad , \tag{2.8}$$

where $\Delta m/S$ is the mass gain per unit area of surface of a specimen; K_p is the rate constant; t is time; A is a constant. The real reaction surface area decreases with time, i.e. is a function of time itself and never reaches the geometric surface area of a specimen. Its value is usually substituted for the value of the real reaction surface area in the equation. It should also be noted that in the case of a complex system of submicron pores (Fig. 2.3), not the diffusion through an oxide layer but the transport of oxygen along the pores to the reaction surface and the removal of interaction products from the pores may become a limiting stage of the process.

Taking into account what was said above, an attempt was made to use a logarithmic equation [2.43] to describe the kinetics of oxidation of porous Si_3N_4 materials:

$$\Delta m/S = K' \lg t + A \tag{2.9}$$

as well as exponential equations with $n \geq 3$ [2.21]

$$(\Delta m/S)^n = K'' t + A \quad , \tag{2.10}$$

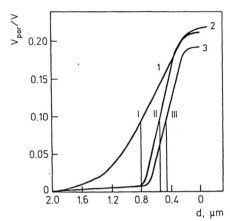

Fig. 2.3. Integral curves of pore size distribution in RBSN of a NKKKM–84 grade: initial state (1) and after oxidation for 4 h at 750° (2) and 900° C (3); d_{50}: $I - 0.8\,\mu m$; $II - 0.54\,\mu m$; $III - 0.44\,\mu m$

which can describe kinetic curves for temperatures at which a liquid oxide layer is formed on the specimen surfaces, or a more general equation of the form

$$\lg(\Delta m/S) = f(\lg t) \quad . \tag{2.11}$$

At lower temperatures the equation for the case of self-blocking pores proposed by *Evans* [2.44] can be used [2.40]:

$$\Delta m/m = (\Delta m/m_\infty)(1 - \exp(-K_a t)) \quad , \tag{2.12}$$

where $\Delta m/m$ can never exceed $\Delta m/m_\infty$ which is the final mass gain that corresponds to the flat section of a kinetic curve. After the formation of a continuous oxide layer on the surface of porous ceramics the situation becomes practically identical to that when the cavities exist near the base of an oxide layer on a dense material, i.e. a new logarithmic equation can be used [2.44]

$$\Delta m/S = K''' \lg(A(t^{1/2} + 1/K)) \quad . \tag{2.13}$$

The asymptotic equation (2.12) describes the initial stages of the process poorly, but better than the parabolic equation (2.8) describes the kinetics of oxidation at long exposure times. When the values of t are relatively small, the asymptotic and logarithmic equations are virtually indistinguishable.

A short analysis of the above kinetic equations demonstrates that the kinetic curves of isothermal oxidation of porous RBSN cannot be described by any one of the equations over the whole studied temperature range and during the total exposure time. This is due to the change of the mechanism of the oxidation process and the pore filling with time and temperature.

Porz and *Thümmler* [2.40] have proposed a quantitative oxidation model of porous silicon nitride taking into account the two concurrent processes occuring in the pore channels: the diffusive flow of oxygen into the pore volume and the consumption of oxygen by its reaction with Si_3N_4 to form SiO_2. The mass gain and penetration depth of oxidation are calculated on the basis of the reaction rates of Si_3N_4 powder, data on oxygen diffusion and the pore characteristics of the RBSN materials, and compared with experimental results. It is clearly shown that, provided the pore channel radii are small, RBSN material may well be used in oxidizing atmospheres without extensive internal oxidation.

In order to study the influence of major additives used in RBSN materials and to choose the optimum one from the point of view of oxidation resistance, the kinetics of ceramics oxidation with ZrO_2, HfO_2, MgO, Y_2O_3 and Al_2O_3, which are widely used in practice were investigated.

The ceramics given in Table 2.2 are produced by injection moulding with further reaction sintering in a nitrogen atmosphere at 1400°–1700° C [2.45].

Table 2.2. Characterization of RBSN ceramics

Additive	HfO_2	ZrO_2	MgO	Y_2O_3	Y_2O_3	Y_2O_3	Y_2O_3
Content,%	2	2	2	2	3.2	5	8.9
Density, kg/m^3	2340	2520	2560	2250	2450	2520	2660

Additive	Al_2O_3	Al_2O_3	Al_2O_3	Al_2O_3
Content,%	2	3.2	6.1	8.9
Density, kg/m^3	2460	2520	2550	2610

Silicon powders (99.9% Si) with a specific surface area of 200–350 m^2/kg and chemically pure magnesia, yttria, alumina, zirconia, hafnia powders with an average particle size of $\sim 1\,\mu m$ are used as raw materials. The milling of powders to prepare a batch results in < 0.6% iron present in ceramics, mainly as silicide Fe_5Si_3. After sintering β-Si_3N_4 becomes a major phase. At the same time α-Si_3N_4 is not revealed. The spectral analysis of ceramics has shown Ti, Cu, Al, Ca, Ni, Mg and Mn in quantities not exceeding 0.01% each [2.27, 45].

The oxidation was investigated at 1000° and 1300° C. Although a large number of similar kinetic curves was plotted, we confine ourselves to the case of a minimum additive content (2%) in our discussion of the results. The exception are ceramics with Y_2O_3 and Al_2O_3, for which more ample data are presented (Fig. 2.4, 5).

At 1000° C the oxidation of the specimens is very pronounced (Fig. 2.5), with its rate decreasing inconsiderably with time, since the porous oxide layer formed on the surface of the specimens has low protective properties. After half an hour of oxidation, the process is well described by the logarithmic equation (2.9).

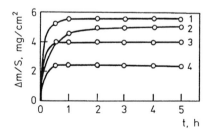

Fig. 2.4. Oxidation kinetics of ceramics with 5% (1); 2% (2); 3.2% (3) and 8.9% Y_2O_3(4) at 1300° C

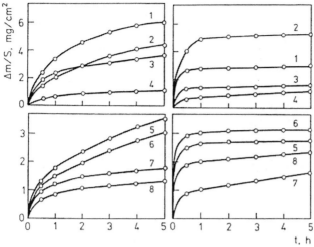

Fig. 2.5. Oxidation kinetics of ceramics with 2% $HfO_2(1)$; $Y_2O_3(2)$; $ZrO_2(3)$; $MgO(4)$ and with 3.2% (5); 2% (6); 8.9% (7) and 6.1% (8) Al_2O_3 at 1000 (a,b) and 1300° C (c,d)

With a certain approximation, the curves for ceramics with Al_2O_3 (Fig. 2.5b) are also described by the parabolic equation; its rate constants calculated by equation (2.8) are shown in Table 2.3.

Table 2.3. Parabolic rate constants $K_p \times 10^4$, $mg^2/(cm^4s)$ for Al_2O_3-doped RBSN

Al_2O_3 content,%	2	3.2	6.1	8.9
1000° C	4.7	6.6	0.8	1.3
1300° C	0.63	0.75	0.9	1.1

Since during oxidation of porous ceramics the reactive surface area changes continuously due to the filling of pores with an oxide phase, its kinetic analysis is very complicated. During the initial period the bulk of the specimen is oxidized, thus the plotted kinetic curves are dependent on the size and shape of specimens. Therefore, the data on the oxidation kinetics of porous materials may be used only as comparative ones and only in those cases when they were obtained on similar specimens. The kinetic parameters calculated using these data are of limited value.

The oxidation at 1300° C results in a liquid silicate layer on the surface of ceramics, therefore the specimens after cooling are covered with a transparent glassy film (Fig. 2.6). X-ray microprobe analysis revealed that the content of calcium, iron and metal of an additive in the film is much higher than their average contents in the specimen. Taking into account that, e.g. the lowest melting temperature is 1660° C for the system SiO_2–Y_2O_3, 1550° C for the system SiO_2–MgO and 1675° C for the system SiO_2–ZrO_2 [2.46], one may suppose that the melting of an oxide layer below 1300° C is caused by the diffusion of impurities to the specimen surface leading to the formation of fusible silicate phases. Under these conditions the content of impurities in the oxide layer

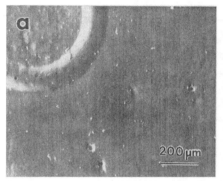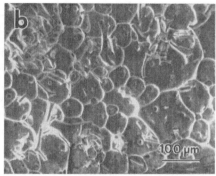

Fig. 2.6. Surface areas of RBSN specimens with 3.2 (a) and 6.1% Al_2O_3 (b) oxidized at 1300° C

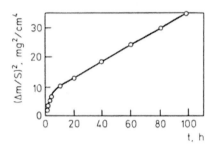

Fig. 2.7. Oxidation kinetics of ceramics with 2% Al_2O_3 at 1000° C

grows with temperature. Thus, after oxidation for 100 h at 1000° C (Fig. 2.7) the concentration of impurities in the surface layer does not reach the level (10%–15%) which is observed after 5 h oxidation at 1300° C. This is explained by the exponential relation between diffusion coefficient and temperature [2.21].

Together with studying the role of additives, in [2.45] an attempt was made to evaluate the effect of porosity on oxidation resistance for the same ceramics. It was found that a liquid silicate layer formed on the specimen surface resulted in an abrupt retardation of oxidation. In this case the mass gain of specimens is mainly determined by oxidation during the initial period, since the oxidation rate is high only untill all the open pores are filled (Fig. 2.8). With an increase

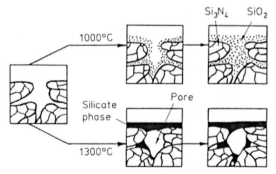

Fig. 2.8. Schematic diagram of oxidation of porous silicon nitride ceramics

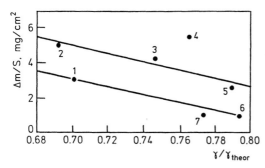

Fig. 2.9. Mass gain of ceramic specimens after 5 h oxidation at 1300° C as a function of relative density of specimens with 2% HfO_2(1); 2%(2); 3.2%(3); 5%(4); 8.9% Y_2O_3(5); 2% MgO(6); 2% ZrO_2(7)

in density of initial specimens their mass gain decreases irrespective of the additive (Fig. 2.9). The obtained curves are close to linear ones, i.e. they obey (2.7). However, Y_2O_3-doped materials are less resistant to oxidation than all the rest. The deviations from linearity for the Y_2O_3-doped ceramics are caused by the fact that the mass gain during the initial stage of oxidation is influenced not only by porosity but also by composition of the intergranular phase and by viscosity and melting temperature of a silicate phase.

At 1000° C the mass gain of a specimen during a certain period of time is dependent on the protective properties of an oxide layer which includes, as follows from XRD data, α-cristobalite, tridymite and, in the case of Y_2O_3-doped materials, also yttrium silicates. It should be noted that an unambiguous dependence between the porosity of specimens and their oxidation resistance at this temperature was not revealed.

The oxidation at 1300° C results in the predominance of α-cristobalite and a glassy phase in the surface layer. It was established that both at 1300° and 1000° C ceramics with MgO appeared to be the most resistant to oxidation, while those with Y_2O_3 and HfO_2 were the least resistant. Higher contents of yttria somewhat increase the oxidation resistance of materials at both temperatures. A ceramic with 5% Y_2O_3 exhibiting low oxidation resistance at 1300° C is an exception. This is apparently due to the presence of $Y_5(SiO_4)_3N$ as a major component of the grain boundary phase. A similar picture is also observed for ceramics with Al_2O_3. It is interesting to note that the oxidation of ceramics with 2% Y_2O_3 is abruptly retarded after an hour of oxidation, while this retardation for ceramics with 3.2% and 5% Y_2O_3 is observed after 0.5 h and for those with 8.9% Y_2O_3 after 20 min. (Fig. 2.5b). This is probably caused by a reduction of viscosity and melting temperature of a silicate phase with increasing yttrium contents [2.46]. It should be noted that an increase in additive contents results in higher contents of a glassy phase in the oxide layer of specimens after their cooling and in lower α-cristobalite and tridymite contents. A content of addivites exceeding 5% leads to the formation of blisters in the surface layer; in the case of Al_2O_3 it is a foamed mass due to the release of gaseous oxidation products (Fig. 2.6b). It is substantial that silicon oxynitride was not revealed either on the surface or in the inner layers of the specimens.

Besides oxide additives, boron is recommended as a sintering aid for RBSN. Its presence facilitates more extensive nitridation of silicon and increases the

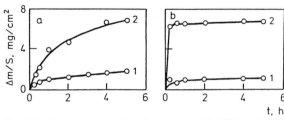

Fig. 2.10. Oxidation kinetics of ceramics with B(1) and BN(2) at 1000° (a) and 1300° C (b)

thermal shock resistance of ceramics. We investigated ceramic specimens with free boron and a graphite-like boron nitride modification added to their batches. B and MgO were added to the first composition (density 2560 kg/m³) and BN to the second one (density 2020 kg/m³). After sintering the diffraction patterns of both specimens revealed α-BN lines. The data demonstrate that the addition of BN did not result in a satisfactory density of the material after sintering, which determined its low oxidation resistance in many respects (Fig. 2.10). At the same time denser ceramics with free boron are only slightly inferior to boron-free ceramics containing only MgO (Fig. 2.5). Judging these results, one should add that a small mass gain for boron-doped ceramics after their oxidation at 1300° C is associated not only with their high oxidation resistance but also with the B_2O_3 vaporization, as it was observed in the case of the B_4C and BN oxidation. It is characteristic that borosilicate glass formed on the specimen surfaces as a result of oxidation at 1300° C flows down the specimens because of the low viscosity and solidifies as beads in their lower part. The phase composition of the oxide layer does not differ from that of the layer formed on the specimens of boron-free ceramics with MgO.

On the basis of these data, one can estimate the effect of additives used for the densification of ceramics on their oxidation.

NKKKM Ceramics. Reaction-bonded Si_3N_4-based ceramics with the addition of MgO exhibits the highest oxidation resistance (Fig. 2.5). However, these ceramics do not possess a very high thermal shock resistance and stability of properties. To get ceramics of higher performance, 30%–40% SiC was added to their batches. Thus, a material with a satisfactory set of mechanical properties could be developed [2.36]. But in order to meet the imposed requirements, certain processing variables were optimized during their development on the basis of the results of mechanical tests and investigations of corrosion resistance. As a result, a whole group of ceramics was obtained that possess a similar chemical composition [2.36] but are somewhat different in their properties (which are given in Table 2.4).

Table 2.4. Characterization of NKKKM ceramics

Ceramics	NKKKM-79	NKKKM-80	NKKKM-81	NKKKM-83	NKKKM-84
Density, kg/m³	2490	2550	2530	2520	2440
V_{uls}, km/s	8.7	8,6	–	8.8	8.1
E_d, GPa	178	198	–	196	160

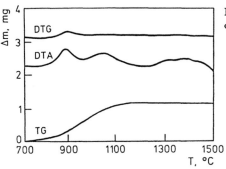

Fig. 2.11. Thermograms of NKKKM-80 oxidation

For producing NKKKM-83–85 ceramics silicon powders of high purity were used, on their basis rather pure Si_3N_4 powders were also prepared. To NKKKM-79 and -80 grades 40% of 30 μm SiC grains were added, for the rest of the grades 30% were used. The MgO content was 1%–2% in all cases. To NKKKM-84 and -85 grades boron was also added. The XRD patterns of sintered specimens of all the modifications revealed β-Si_3N_4 and SiC lines.

Since NKKKM-79–83 ceramics were designed to work below 1400° C and NKKKM-84 and -85 grades below 1200° C, these materials had to meet different requirements which also were taken into account in the analysis of the obtained data. The investigations of ceramics have demonstrated that the general mechanisms of their oxidation are quite similar for different modifications. Therefore, only the most characteristic data or those which are necessary to support the corresponding conclusions are presented as an example to describe the results.

The analysis of thermograms for the ceramics[1] (Fig. 2.11) has shown that their oxidation starts at temperatures close to 800° C. Here the first peak on the differential thermal analysis (DTA) curve corresponds to the oxidation of Si_3N_4 (indicated as lower intensity of nitride lines on the diffraction patterns after oxidation at 800°–900° C), the second one probably represents the SiC oxidation (after heating up to 1100° C the relative intensity of carbide lines decreases). The predominant oxidation of Si_3N_4 in the surface layer is also confirmed by the petrographic analysis. The oxidation of ceramics in air leads to the formation of silica (Fig. 2.12a), the heating of specimens in a closed space with an oxygen-depleted atmosphere results in formation of silicon oxynitride on their surface (Fig. 2.12b). The petrographic analysis of the surface layer of specimens oxidized in air below 1200° C also revealed 10%–15% of amorphous magnesium silicates. Their crystallization could be reached only on very slow cooling of the specimens.

---

[1] On thermograms, as opposed to kinetic curves, not a value of mass variation per geometric unit surface area but a value of mass variation recorded by an instrument on 3.5 × 5 × 20 mm specimens is presented in the majority of cases, since TG-curves were used only to estimate qualitatively the obtained results and to determine the initial temperature of each oxidation stage. They were never used to calculate kinetic parameters.

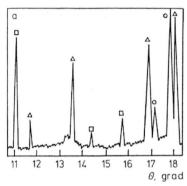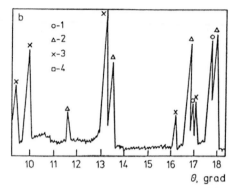

Fig. 2.12. Diffraction pattern of NKKKM-79 specimens oxidized in air at 1200° C (a) and in a closed silicon nitride chamber at 1450° C (b): SiC(1); β-Si$_3$N$_4$(2); Si$_2$N$_2$O(3); SiO$_2$ (α-cristobalite)(4)

The thermogravimetric (TG) curve (Fig. 2.11) shows that at 1100° C the mass gain of the specimens is abruptly retarded, which is caused by the melting of an oxide layer [2.43]. The weak thermal effects on the DTA curve at temperatures exceeding 1300° C are apparently due to the transformations in the oxide layer. The oxidation of silicon nitride ceramics as well as the oxidation of NKKKM ceramics causes the diffusion of impurities exhibiting a high affinity for oxygen to the specimen surface. In this case the areas with iron inclusions in the surface layer give rise to pore formation (Fig. 2.13a). This is connected with the fact that iron silicates possess the lowest melting temperature compared with other silicates formed under these conditions [2.46], and the bubbles of gaseous oxidation products probably break through the oxide layer in its weakest points. The studies on the oxidation kinetics of NKKKM ceramics under isothermal conditions confirmed the results of the analysis of oxidation under non-isothermal conditions (Fig. 2.14).

Despite a similar run of kinetic curves for the initial stages of the process, the studied modifications differ in their oxidation resistance, viz. the lowest mass gain after 5 h oxidation is observed for NKKKM-83 ceramics, and the highest one for NKKKM-81. The content of impurities is apparently a decisive factor. NKKKM-83 ceramics are characterized by a minimum content of impurities while NKKKM-81 contains maximum quantities of impurities. The content of impurities in ceramics also determines the melting point and the amount of silicate phase formed on the specimen surface. Thus, NKKKM-79–81 ceramics after oxidation were covered with a glassy film already at 1100° C (Fig. 2.13), while the same film was formed on NKKKM-83 ceramics only at 1300° C. On NKKKM-84 and -85 the glassy film was formed already at 900° C due to boron present in their composition, but because of lower viscosity this layer exhibited poorer protective properties than the layer formed on boron-free ceramics specimens. When the layer was formed above 1200° C, the oxidation rate for NKKKM-84 and -85 was much higher than for other modifications. Therefore these ceramic modifications can be used only for components operating below 1200° C.

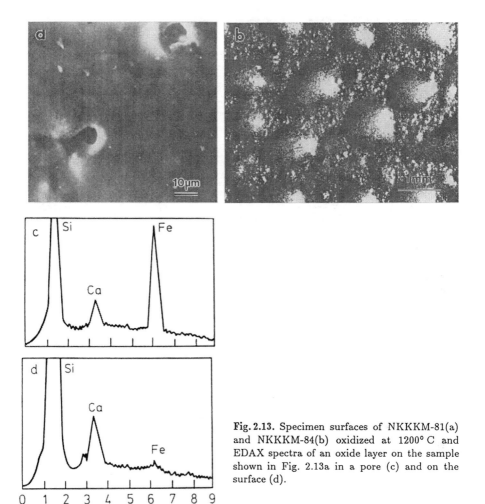

Fig. 2.13. Specimen surfaces of NKKKM-81(a) and NKKKM-84(b) oxidized at 1200° C and EDAX spectra of an oxide layer on the sample shown in Fig. 2.13a in a pore (c) and on the surface (d).

The data presented in Fig. 2.14 demonstrate that the kinetic curves obey (2.9). The curves in Fig. 2.15a obtained for the specimens oxidized in air over a long period of time at 800° C are relatively well described by (2.8). However, if we present these data as plots of mass gain squared against oxidation time (Fig. 2.15b), their bends give evidence of a change in the mechanism of the process with time. Thus, on the curve for NKKKM-81 ceramics at 1000° C horizontal sections were observed after 10% h oxidation because of the formation of a protective glassy film on the specimen surfaces. On the specimens of other ceramic modifications a silicate layer was not yet formed at these temperatures and the bends on their curves can be attributed to the filling of pores with oxidation products. However, even after 100 h exposure to 800° C the oxidation rate (NKKKM-81) decreases only inconsiderably.

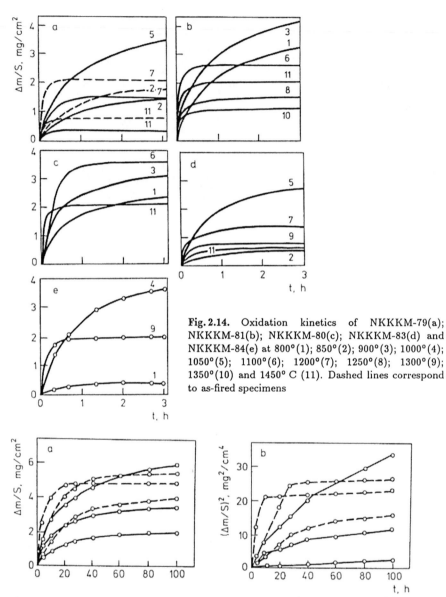

Fig. 2.14. Oxidation kinetics of NKKKM-79(a); NKKKM-81(b); NKKKM-80(c); NKKKM-83(d) and NKKKM-84(e) at 800°(1); 850°(2); 900°(3); 1000°(4); 1050°(5); 1100°(6); 1200°(7); 1250°(8); 1300°(9); 1350°(10) and 1450° C (11). Dashed lines correspond to as-fired specimens

Fig. 2.15. Oxidation kinetics of NKKKM-81(1), NKKKM-79(2) and NKKKM-83(3) at 800° C (solid lines) and at 1000° C (dashed lines) in $\Delta m/S - t$ (a) and $(\Delta m/S)^2 - t$ (b) coordinates

2.1.5 Oxidation of Dense Ceramics

Dense silicon nitride ceramics are produced by hot pressing (HP) and hot isostatic pressing (HIP) as well as by gas-pressure or pressureless sintering. As is seen in Fig. 2.16, the oxidation resistance in air for ceramics that were hot-pressed in graphite molds and ceramics that were sintered under nitrogen pres-

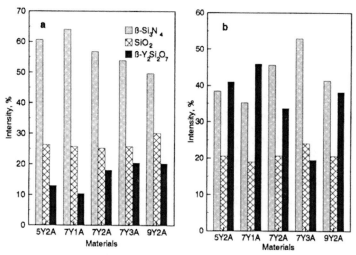

Fig. 2.16. Semiquantitative XRD analysis of the surface layer of HPSN (a) and SSN (b) samples oxidized for 100 h at 1200° C in air

sure differs but not very much. This may be explained by the variations in the phase composition of specimens prepared from similar batches but by different technologies. More often, however, the differences in oxidation resistance of sintered and hot-pressed materials are connected with a higher porosity of the former and a higher content of oxide additives in them.

The role of additives used as sintering aids was studied by many researchers [e.g. 2.47, 48]. It was shown that the oxidation of the grain-boundary phase in ceramics may start at lower temperatures than the oxidation of Si_3N_4 itself (Fig. 2.17). This is due to the formation of zirconium oxynitride phases [2.49], Y-apatite [2.48], Nd-apatite, melilite or similar cerium, lanthanum, samarium oxynitrides and others [2.50] along the grain boundaries. Babini et al. [2.47] developed a diffusion model of the oxidation of hot-pressed materials in the system $Si_3N_4 - Y_2O_3 - SiO_2$ in which the most relevant parameters are the amount and composition of the grain-boundary phase, the width of the diffusion zone and the concentration gradient at the Si_3N_4/oxide reaction interface. The model accounts for both kinetic (oxidation rate constants) and thermodynamic (apparent activation energy for oxidation) variations and proves suitable for the application to other additive systems. It is suggested that a more appropriate evaluation of the thermodynamic parameters of the diffusion process must account for the variation of concentration profiles of the diffusing species with temperature.

Another model based on the schematic diagram in Fig. 2.17 was proposed by the authors of [2.48]. It describes well the oxidation kinetics of HPSN with 9% Y_2O_3. It should be noted that the oxidation of ceramics containing rare-earth oxides leads, as a rule, to the formation of disilicates $Me_2Si_2O_7$ [2.50].

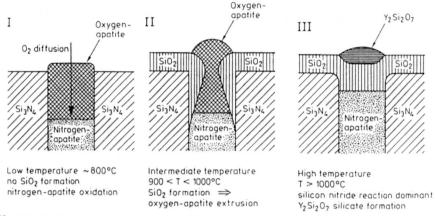

I   Oxygen-   II   Oxygen-   III $Y_2Si_2O_7$
apatite   apatite

O_2 diffusion

Low temperature ~800°C   Intermediate temperature   High temperature
no SiO_2 formation   900 < T < 1000°C   T > 1000°C
nitrogen-apatite oxidation SiO_2 formation $\Rightarrow$ silicon nitride reaction dominant
                               oxygen-apatite extrusion $Y_2Si_2O_7$ silicate formation

Fig. 2.17. Oxidation model for dense Y-containing Si_3N_4 ceramics [2.48]

The kinetic curves of oxidation of hot-pressed materials (Fig. 2.18) are usually described by (2.8). But when a porous oxide layer is formed on the specimen surface, the mass gain becomes linear with time [2.51].

An abrupt increase in activation energy of HPSN oxidation above 1350° C (Table 2.5) has led to the assumption [2.55] that although parabolic oxidation kinetics was observed over the whole temperature range, oxygen diffusion must have predominated at low temperatures, whereas at $T \sim 1350°$ C metal cation diffusion through the grain-boundary phase seemed to be the limiting factor.

The specimens of HPSN with a density close to the theoretical one are mainly oxidized on the surface. Therefore not an increase, as in the case of RBSN oxidation, but a certain decrease of their density after oxidation is ob-

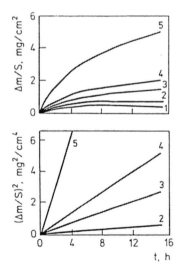

Fig. 2.18. Oxidation kinetics in oxygen for a hot-pressed material HS–130 (Norton Co.) in $\Delta m/S - t$ (a) and $(\Delta m/S)^2 - t$ (b) coordinates: 1090°(1); 1205°(2); 1260°(3); 1315°(4); 1370° C (5)

Table 2.5. Activation energies (E) of oxidation of silicon nitride powders and several Si_3N_4 materials

Material	Atmosphere	T,° C	E, kJ/mole	References
β-Si_3N_4 powder	air	1000–1100	280	[2.52]
β-Si_3N_4 powder	air	1200–1500	390	[2.52]
Si_3N_4 powder	dry air	1065–1340	285	[2.24]
Si_3N_4 powder	dry oxygen	1065–1340	255	[2.24]
RBSN	air	1100–1200	130	[2.53]
RBSN	air	< 1100	320	[2.53]
HPSN + 1% MgO	dry oxygen	1000–1400	255	[2.15]
HPSN + 1% MgO	humid oxygen	1200–1400	375	[2.30]
HPSN + CeO_2 + SiO_2	air	1100–1370	350	[2.54]
HPSN + 8% Y_2O_3 + 1% MgO	air	< 1150	120	[2.23]
HPSN + 8% Y_2O_3 + 1% MgO	air	1150–1350	580	[2.23]
HPSN + 8% Y_2O_3 + 1% MgO	air	> 1350	960	[2.23]
Hot–pressed $Si_{2.9}Be_{0.1}N_{3.8}O_{0.2}$	air	1400–1500	700	[2.38]
CVD Si_3N_4 (crystalline)	dry oxygen	1550–1650	390	[2.35]
CVD Si_3N_4 (amorphous)	dry oxygen	1550–1650	460	[2.35]
HPSN + Y_2O_3 + SiO_2	air	1150–1400	260–623	[2.47]
HPSN + 5%Al_2O_3 + 12%ZrO_2	air	< 1200	100	[2.55]
HPSN + 5%Al_2O_3 + 12%ZrO_2	air	> 1200	800	[2.55]
HPSN + 10%BN + 5%MgO	air	1200–1400	650	[2.56]

served. An increase in the sizes of the specimens after oxidation was reported by the authors of [2.49, 57].

Apart from oxide additives, other refractories exerting considerable influence on the oxidation resistance of ceramics are added to improve their operating characteristics. Thus, boron nitride is added to increase thermal shock resistance and to improve the dielectric characteristics. We studied the oxidation of HPSN with 10% BN and 5% MgO in air at temperatures below 1500° C [2.56]. The XRD analysis revealed β-Si_3N_4, Mg_2SiO_4, $MgSiN_2$, BN and an amorphous phase in the composition of the initial specimens.

The studies on oxidation under non-isothermal conditions (Fig. 2.19) at temperatures up to 1000° C demonstrated a small mass loss due to the desorption of adsorbed substances and the vaporization of boron oxide that was present on the surface of the initial specimens or formed on the BN oxidation starting from temperatures close to 900° C [2.58]. The oxidation of Si_3N_4 starts at temperatures of about 800° C, but a detectable mass gain for HPSN is usually observed at temperatures close to 1000° C [2.51]. The mass gain of the

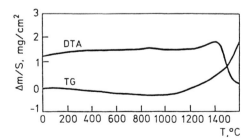

Fig. 2.19. Thermograms of HPSN-BN oxidation

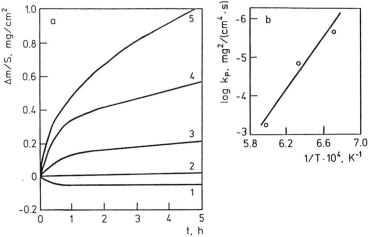

Fig. 2.20. Oxidation kinetics of HPSN-BN at 800°(1); 1000°(2); 1200°(3); 1300°(4) and 1400° C (5), and rate constant logarithm as a function of temperature (b)

ceramic composite specimens is also detected at temperatures above 1000° C and is accompanied by a very weak exothermal peak on the DTA curve. However, since the experiment was performed under non-isothermal nonequilibrium conditions, all the processes shifted to higher temperatures. At temperatures above 1200° C the oxidation process becomes more rapid. It is accompanied by a noticeable thermal effect (Fig. 2.19). At temperatures close to 1400° C the rate of mass gain grows, but the exothermal effect changes into an endothermal one connected with an abrupt increase in the vaporization rate of boron oxide [2.58]. Later on, while heating up to 1500° C (Fig. 2.19) and even up to 1600° C the mechanism of the process does not change.

The results of studying the oxidation kinetics under isothermal conditions (Fig. 2.20) confirm the data obtained during programmed heating of specimens. The kinetic curves of oxidation in the range of 1200°–1400° C obey the parabolic law (2.8), since in the mentioned temperature range the oxidation rate is limited by the diffusion of elements through the oxide layer. The value of the apparent activation energy of oxidation 650 kJ/mol was determined by the Arrhenius equation according to the plot presented in Fig. 2.20b. The obtained value of E is close to the data found in the literature for the oxidation of other HPSN materials in air (Table 2.5).

The XRD analysis of the specimens exposed to 800°–1000° C for 5 h did not reveal any changes in the phase composition, though a white deposit formed on their surface was observed. Oxidation at 1200°–1500° C results in the formation of α-cristobalite, forsterite, enstatite and borosilicate glass. In this case up to 1300° C the oxidation products solidify after cooling as a smooth glassy film uniformly covering the specimen surface. At higher temperatures separate areas of the surface become blistered. After oxidation at 1500° C the surface of the specimens is rough, with a great number of blisters and craters; the traces of active B_2O_3 vaporization and evolution of gaseous products through the oxide layer as a result of Si_3N_4 and BN oxidation clearly visible. The diffraction patterns of the specimens heated up to 1500° C do not reveal $MgSiN_2$ lines.

Thus, the oxidation of the studied ceramics at temperatures above 1200° C proceeds by reactions (2.1), (2.5), (2.6) as well as by the reactions $2\,MgSiN_2 + 3\,O_2 = 2\,MgSiO_3 + 2\,N_2$ and $4\,BN + 3\,O_2 = 2\,B_2O_3 + 2\,N_2$. The presence of B_2O_3 in the oxide layer decreases its viscosity and melting temperature [2.46], which determines the slightly lower oxidation resistance of the studied ceramics as compared to BN-free MgO-doped hot-pressed materials (Fig. 2.18).

It is also necessary to take into account that the ceramics contained rather large amounts of MgO, and an increase in magnesia contents decreases the oxidation resistance.

Thus, BN- and MgO-doped hot-pressed ceramics exhibit high oxidation resistance in air up to 1200° C. At higher temperatures the process becomes more rapid and above 1400° C intensive oxidation of ceramics starts. Therefore, these ceramics cannot be applied in the mentioned temperature range.

As an example, we also give some information on the oxidation behaviour of sintered ceramics containing yttria and alumina as sintering aids. The XRD analysis of the material revealed β-Si_3N_4, traces of carbide and tungsten silicides W_5Si_3 and WSi_2. In addition to the above elements, the X-ray microprobe analysis revealed K, Ca, Fe, Zr and Mo impurities, with their content being $\sim 0.01\%$ each.

Scanning electron microscopy (SEM) investigations of ceramic surfaces have demonstrated that tungsten compounds are present as point inclusions of up to $2\,\mu m$ in size. The porosity of the specimens did not exceed 5%, and the average pore size was $\sim 0.1\,\mu m$.

As is seen in Fig. 2.21, the mass variation of sintered ceramic specimens due to oxidation starts at temperatures above 1100° C and is accompanied by a very weak exothermal effect. Above 1200° C the oxidation process becomes more rapid and up to 1500° C the TG curve demonstrates the mass gain due to the Si_3N_4 oxidation by reaction (2.1). At the same time above 1300° C the DTA curve shows a well-pronounced endothermal peak with a maximum at 1500° C corresponding to the vaporization of one of the oxidation products.

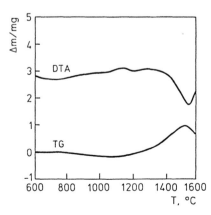

Fig. 2.21. Thermograms of oxidation of sintered ceramic ($Si_3N_4 + Y_2O_3 + Al_2O_3$)

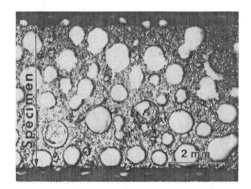

Fig. 2.22. Surface of a SSN specimen after heating up to 1570° C at a rate of 15° C/min

Above 1500° C the vaporization rate exceeds the rate of formation of condensed oxidation products, and the mass gain changes into a mass loss on the TG curve. The mass of specimens decreases, e.g., as a result of the WC oxidation by the reaction $2\,WC + 5\,O_2 = 2\,WO_3 + 2\,CO_2$. Volatile tungsten oxide vaporizes from the specimen surface. The mass loss may also be due to the interaction of the oxide layer with Si_3N_4 by reaction (2.4). The morphology of the specimen surface after the heating up to 1570° C is presented in Fig. 2.22. White blisters are formed on the ceramic surface due to the evolution of gaseous oxidation products at 1400° C and more, with the size of these blisters increasing with temperature. An X-ray microprobe analysis established that the silicate phase forming these blisters was Ca-, Y-, Al- and Mo-enriched.

Thus, because of the vaporization of oxidation products and the absence of a continuous protective film, the performance of ceramics above 1300° C cannot be high. At the same time these ceramics, as opposed to RBSN, exhibit a rather high oxidation resistance at temperatures $\sim 1000°$ C. On the whole, these results confirm a suitability of such materials for manufacturing e.g. the components of Diesel engines operating at 900°–1100° C [2.59]. The oxidation resistance of sintered ceramics can be increased by crystallization of the grain boundary phase (Fig. 2.23). The isothermal exposure of the specimens to 1400°–1450° C for 5 h [2.61] results in the transition of the amorphous secondary phase to the crystalline state.

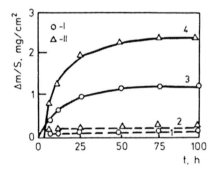

Fig. 2.23. Oxidation kinetics of sintered ceramics ($Si_3N_4 + Y_2O_3 + MgO$) with the crystalline (I) and the amorphous (II) secondary phases at 1000° (1,2) and 1300° C (3,4) [2.60]

Among silicon nitride materials sialons are often considered as a separate group. It is characteristic of them that the elements of an additive (Al and oxygen in the case of β-sialons or Y, Al and oxygen in the case of α-sialons) dissolve in the Si_3N_4 lattice, which leads to a change in the chemical and physico-mechanical properties of the ceramics [2.37]. Recently, solid solution-based materials with a β- or α-Si_3N_4 lattice containing Mg, Be and other elements have also been developed. The oxidation resistance of sialon ceramics is discussed in a number of papers [e.g. 2.62–64]. It is reported in [2.63] that the presence of residual β-Si_3N_4 and a glassy phase decreases the oxidation resistance of sialon ceramics. The kinetic curves are also described by the parabolic law, the composition of the oxidation products does not differ from that of the products found on the surface of other silicon nitride materials containing Al_2O_3 or $Al_2O_3 + Y_2O_3$. Thus, one can conclude that the oxidation behaviour of sialon ceramics is not so different from that of other Si_3N_4 materials. At the same time single-phase sialon materials which do not contain the glassy phase or contain a minimum amount of the grain-boundary phase possess very good oxidation resistance even above 1400° C [2.65].

Among dense silicon nitride materials CVD Si_3N_4 occupies an isolated position. Materials produced by chemical vapour deposition [2.35] do not contain sintering aids and are superior in oxidation resistance to hot-pressed and sintered ones (Chap. 7).

2.1.6 Peculiarities of Oxidation Kinetics

All the above kinetic curves for silicon nitride ceramics gave evidence of the retardation of oxidation with time, thus of promising potentials for the application of such materials at high temperatures in oxidizing environments. However, the precise investigations of the oxidation kinetics demonstrate that in some cases after a certain period of time the oxidation process accelerates (Fig. 2.24). And a similar phenomenon was revealed both in the range of intermediate and high temperatures, for dense and porous materials with different additives at exposure times from a few tens of minutes to a few tens of hours. Such a change in oxidation kinetics was demonstrated by thermogravimetric analysis [2.49, 66] and by the analysis of gaseous oxidation products [2.13]. It should be noted that many researchers recorded similar curves, but such behaviour of ceramics was either not discussed at all [2.67], or the bend on the curve was ignored and the curve was reduced to a conventional parabola [2.68]. When the oxidation kinetics is studied by periodic weighing, it is virtually impossible to notice such a change. Taking the above into account, one may assume that such an instability of the oxidation process in time is inherent in many materials. What are the reasons? Until recently the following assumptions have been expressed:

1) change of composition of ceramics during longer oxidation [2.49];
2) crystallization of a glassy oxide layer [2.69];
3) failure (cracking, spalling) of the oxide layer [2.13,66];
4) acceleration of the oxidation process due to internal stresses [2.66].

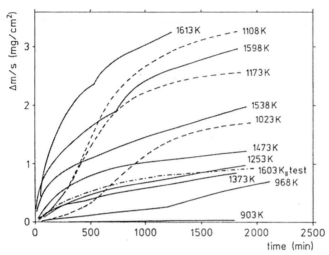

Fig. 2.24. Oxidation kinetics curves of HPSN with 9.7 mol% MgO, 3.44 mol% Y_2O_3 and 22.91 mol% ZrO_2 [2.49]

As a fifth possible reason one should assume the variations in the composition of the oxide layer due to the diffusion of additives and impurities to the surface and solid-phase and liquid-phase reactions proceeding in the oxide layer, which changes its protective properties.

At the same time all these mechanisms cannot develop simultaneously and over the whole temperature range. The cracking of the oxide layer can explain the acceleration of oxidation at temperatures up to ~1000° C, but not at higher temperatures when the relaxation of internal stresses occurs easily due to the softening of the oxide layer and the grain-boundary phase. The variation of the composition of ceramics or the oxide layer due to diffusion processes can hardly lead to such an abrupt uneven variation of the oxidation rate (Fig. 2.24). The authors of the cited works did not succeed in finding unambiguous evidence of other mechanisms of acceleration of oxidation. This problem requires additional efforts, since the acceleration of oxidation of ceramic components under operating conditions may limit their applicability.

2.2 Silicon Carbide Ceramics

Silicon carbide materials found industrial application much earlier than Si_3N_4-, AlN-, B_4C- and BN-based materials. Silicon carbide refractories with a silica additive (90% SiC+10% SiO_2) were used as early as in the 1920s, in the 1950s silicon carbide with a silicon nitride additive (75% SiC+25% Si_3N_4) was employed for manufacturing the nozzles of rockets like NIKE and BOMARC [2.70]. Therefore studies on the SiC oxidation were started earlier than investigations of other ceramic materials. The monographs devoted to silicon carbide and silicon carbide materials [2.71, 72] include data on the thermodynamics of processes taking place during the interaction of SiC with oxygen, on the oxidation kinetics of powders, single crystals and polycrystalline materials. However, the salient features of the oxidation of different silicon carbide materials developed and presently produced were not examined in the above monographs. So in the following we will concentrate on describing the oxidation process of advanced materials produced by different technologies.

2.2.1 Thermodynamics of SiC Oxidation

Thermodynamic calculations [2.71] show that as a result of SiC oxidation the formation of SiO_2 is the most probable, and it proceeds according to the reactions

$$SiC + 2O_2 = SiO_2 + CO_2 \quad , \tag{2.14}$$

$$2SiC + 3O_2 = 2SiO_2 + 2CO \quad . \tag{2.15}$$

The thermodynamic probability of the reaction

$$SiC + O_2 = SiO + CO \tag{2.16}$$

leading to the formation of SiO, as in the case of Si_3N_4, increases with growing temperature and decreasing partial oxygen pressure. Figure 2.25 presents the plot showing the areas of active oxidation with the formation of only gaseous products and the areas of passive oxidation accompanied by the SiO_2 formation. As is reported in [2.74], the temperature of active/passive oxidation transition is also dependent on the gas flow rate.

The silica layer formed on the specimen surface exhibits protective properties up to 1630° C [2.71]. At higher temperatures the interaction of the materials with the oxide layer starts by the reaction

$$SiC + 2SiO_2 = 3SiO + CO \quad . \tag{2.17}$$

2.2.2 Composition of Oxide Layer

The studies on the oxidation products of powders and single crystals at temperatures up to 1400° C [2.75] demonstrate that amorphous silica is formed in all cases, but under certain conditions the formation of cristobalite is also ob-

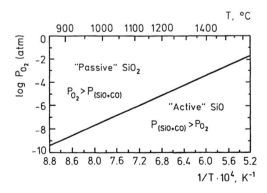

Fig. 2.25. Partial oxygen pressures on the active/passive oxidation transition for SiC [2.73]

served. On oxidation of $\alpha - SiC$ single crystals cristobalite is not formed up to 1400° C. At the same time the oxidation at 1300° C for over 5 h of crystals that are a mixture of polytypes results in the appearance of characteristic cristobalite rosettes on the surface. The formation of cristobalite is observed already at 1100° C on the oxidation of SiC powders of different grain sizes [2.75] and materials containing, apart from SiC, alumina and graphite [2.76]. However, in the temperature range of 1100°–1300° C cristobalite is actively formed only after 20–30 min of oxidation, i.e. a certain incubation period exists for cristobalite nuclei to form in the layer of amorphous SiO_2. At 1400° C cristobalite was formed from the first minutes of oxidation [2.75].

The authors of [2.69], who investigated the kinetics of CO_2 formation on the oxidation of SiC powders, consider that cristobalite develops at temperatures above 870° C after the exposure to an oxidizing environment for 100–200 min, the cristobalization of an oxide layer is accompanied by a bend on the kinetic curves of oxidation (Fig. 2.26). As in the case of silicon nitride ceramics (Fig. 2.24), the period of time until oxidation begins to accelerate decreases with temperature. The formation of cristobalite in the oxide layer is also accompanied by an increase in activation energy (Table 2.6), from the value typical of silicon single crystals to the values closer to those of silicon carbide ceramics. It is interesting to note that, in opinion of the authors of [2.69], the SiC oxidation starts already at 300° C and proceeds differently in different temperature ranges (Fig. 2.26).

The analysis of literature data in [2.79] demonstrates that the activation energy of the oxidation of powders, fibres and SiC single crystals in oxygen and air lies in very wide ranges, from 80 to 630 kJ/mole, with these values increasing with temperature (also Table 2.6).

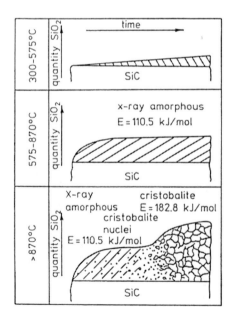

Fig. 2.26. Model representation of silicon carbide oxidation [2.69]

Table 2.6. Activation energies of SiC oxidation

Material	Range of activation energies, kJ/mole	Temperature range, ° C	References
Single crystal silicon	120	1200–1400	[2.77]
SiC powder	110	575–870	[2.69]
	183	870–1300	[2.69]
Single crystal SiC			
Black, fast growth face	134–197	1200–1500	[2.77]
Green, fast growth face	121–297	1200–1500	[2.77]
Black, slow growth face	372	1200–1500	[2.77]
Green, slow growth face	339	1200–1500	[2.77]
CVD SiC, CNTD	142–293	1200–1500	[2.77]
Sintered α-SiC	217–289	1200–1450	[2.77]
HPSiC	221	1200–1400	[2.77]
HPSiC + 4%Al_2O_3	485	1200–1400	[2.78]
Self-bonded SiC	480	1100–1400	Authors' data

There are contradictory opinions whether silicon oxycarbides appear on SiC oxidation. The assumptions of the existence of compounds of SiC_2O, SiC_3O, $SiCO_3$, SiCO types called siloxicons were suggested already in the 1930s. But later investigations showed that these compounds are fine SiO_2, SiC and carbon mixtures. By the beginning of the 1970s the opinion was formed that silicon oxycarbides did not exist [2.80]. This was based on the results of XRD and chemical analyses. X-ray photoelectron spectroscopy (XPS) used to analyse the energy of a silicon bond proved the possibility of the existence oxycarbides [2.81]. Unstable compounds of SiCO type were revealed in oxidation products of chemical-vapour deposited (CVD) SiC by IR and XPS. They were formed as intermediate products at the SiC/SiO_2 interface and at the initial stage of oxidation at temperatures of $\sim$ 1200° C. The phase revealed in vacuum is thermally unstable and decomposes into silicon, SiO_2 and CO (CO was detected by mass-spectrometry) at 1100°–1200° C.

2.2.3 Hot-Pressed and Sintered SiC

The oxidation process of HPSiC obeys the parabolic law (2.8) and, as in the case of Si_3N_4, is largely dependent on the kind and amount of sintering aids and the content of impurities. In the majority of cases Al_2O_3 is used as a sintering aid for SiC. It is noted in [2.82] that a higher content of Al_2O_3 reduces the oxidation resistance of ceramics (Fig. 2.27) and influences the rate constant of the parabolic oxidation K_p, mg^2/(cm^{4}s) of HPSiC in dry oxygen at 1370° C. The data are shown in Table 2.7.

Table 2.7. Parabolic rate constants (K_p) for Al_2O_3-doped SiC [2.82]

Al_2O_3 content, %					
in a batch	2.5	6.1	9.2	12.1	15.0
after hot pressing	1.8	5.7	–	11.0	–
$K_p \cdot 10^6$, mg^2/(cm^{4}s)	1.94	2.77	4.17	6.82	7.78

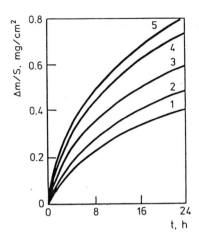

Fig. 2.27. Oxidation kinetics in dry oxygen at 1370° C for HPSiC with 2%(1); 6.1%(2); 9.2%(3); 12.1%(4) and 15% Al_2O_3(5) [2.82]

This is due to the diffusion of aluminium to the surface and its dissolution in the oxide layer, which is accompanied by a deterioration of the protective properties of the layer and a higher gas permeability. The formation of a liquid phase is clearly seen in the areas of large Al_2O_3 inclusions. In the oxide layer after oxidation at 1370° C cristobalite and a small amount of mullite were revealed. On oxidation of composite materials containing SiC and α-Al_2O_3 [2.83] mullite was shown to form above 1300° C.

The kinetics of oxidation of HPSiC with 4% Al_2O_3 containing also 4% WC and Al, Fe, Ti, Ca, Cr, Ni, Mg, Na and K impurities (in decreasing order from 1 to 0.001%) was studied in oxygen at atmospheric pressure and temperatures of 1200°–1400° C [2.78]. The kinetic curves also obeyed the parabolic law, but only after the second hour of heating. Because of volatile tungsten oxide, the rate of mass gain during the first 2 h was lower. The apparent activation energy of oxidation of this material in the mentioned temperature range was 485 kJ/mole (Table 2.6). The oxide layer contained cristobalite and a glassy phase in which aluminium, iron, potassium, sodium and other impurities, tungsten excluded, were concentrated. The content of tungsten in the material is higher than in the oxide layer; in view of this the author of [2.78] concluded that the tungsten oxide vaporizes at the initial stage of oxidation. The removal of CO from the SiC/SiO_2 interface is considered to be a limiting stage of a steady-state oxidation process.

The fact that the kinetic curves obey the parabolic law shows that the oxidation of SiC as well as that of Si_3N_4 is described by the Wagner theory. What is more, the electrochemical theory of oxidation is experimentally corroborated using the system $Si-SiO_2$ [2.71] close to the investigated composition. Studying the effect of the electric field on the oxidation of silicon, Jorgensen has demonstrated that depending on the polarity the external electric field accelerates or retards the diffusion of oxygen ions in the SiO_2 layer and even entirely closes the galvanic cell $Si|SiO_2|O_2$, thus terminating the oxidation process at a given temperature [2.71]. The parabolic growth of a SiO_2 film on SiC apparently follows a similar mechanism, but is complicated by a mutual diffusion of gaseous carbon oxides.

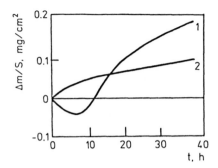

Fig. 2.28. Oxidation kinetics of sintered α-SiC with 1% B (1) and 0.8% Al (2) at 1200° C [2.84]

In [2.84] the oxidation of dense (porosity lower than 2%) sintered α-SiC containing up to 0.8% Al and up to 1% B was investigated at temperatures of 900°–1600° C. The addition of boron (Fig. 2.28) leads to the mass loss of specimens at the initial stage of oxidation. The kinetic curves of oxidation of Al-doped materials or materials with a mixture of 0.3% Al and 0.4% B obey the parabolic law at the beginning of the process. The oxidation of B-doped materials at 1200° C for 200 h results in the formation of an oxide layer consisting of 80%–90% SiO_2 and 10%–20% B_2O_3 on their surface. As a result of oxidation of Al-doped sintered materials as well as hot-pressed materials with Al_2O_3, the oxidation products apart from silica, included mullite.

2.2.4 Self-Bonded and Recrystallized SiC

The oxidation process for self-bonded SiC is similar to the oxidation of silicon. Self-bonded SiC is a two-phase material which contains, apart from carbide, $\sim 10\%$ of free silicon. A spectral analysis of the material produced by the Brovary Powder Metallurgy Plant revealed Ca, Fe, Mg, Al, Sn, Ti ($\sim 0.1\%$ each) and Cu, Ni, Cr, Mn ($\sim 0.01\%$ each) impurities in its specimens.

The oxidation of self-bonded SiC (Fig. 2.29a) starts above 1000° C. On the DTA curve the first (broad) exothermal peak is observed in the temperature range of 1000°–1400° C. The corresponding TG curve shows that the effect of

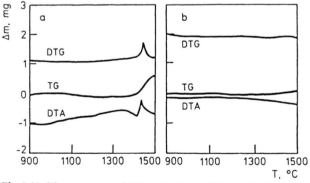

Fig. 2.29. Thermograms of SiC oxidation: self-bonded (a) and recrystallized of AnnaNox CK grade (b)

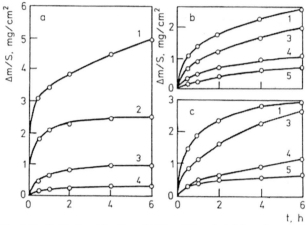

Fig. 2.30. Oxidation kinetics of SiC: self-bonded (a) and recrystallized of AnnaNox CK (b) and Crystar (c) grades at $1400°(1)$, $1300°(2)$, $1200°(3)$, $1100°(4)$ and $1000°$ C(5)

mass variation in this area is very small. The second (narrow) exothermal peak on the DTA curve is observed at higher temperatures and is accompanied by a maximum on the DTG curve. In the range of $1400°–1500°$ C the mass gain of specimens increases considerably, because SiC is intensively oxidized to SiO_2 by reactions (2.14, 15) and the oxidation of silicon proceeds in accordance with the reaction $Si + O_2 = SiO_2$.

The investigations of the oxidation process of ceramics under isothermal conditions (Fig. 2.30) have shown that the kinetic curves obey well the parabolic law (2.8). As in the investigations under nonisothermal conditions, mass variations of self-bonded SiC specimens as a result of heating at $1000°$ C were not observed. The absence of oxidation at temperatures up to $1100°$ C is also confirmed by electron microscopy data. In particular, on the specimen kept at $1050°$ C for 3h there was no oxide layer and the texture from mechanical pretreatment was well visible [2.85]. The morphology of the specimen surface oxidized at $1350°$ C is shown in Fig. 2.31. The surface is covered with large pores and craters apparently developed due to gas evolution through the oxide layer. The micrograph also shows a drop formed on the melting of a silicon inclusion.

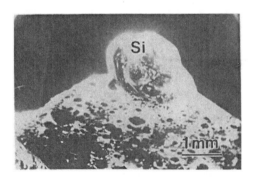

Fig. 2.31. Surface of a self-bonded SiC specimen oxidized for 3h at $1350°$ C

To elucidate the effect of factors like porosity and impurity contents on the oxidation resistance of silicon carbide ceramics, the experiments with recrystallized SiC of Crystar grade (Norton Co., USA) and of AnnaNox CK grade (AnnaWerk, Germany) were performed under similar conditions.

The total content of metal impurities was $\sim 0.1\%$ and of oxygen $\sim 1\%$. These materials are manufactured as plates with a thickness of 6 mm used for ceramic firing. Since the structure of recrystallized SiC varies across the thickness of the plate, test specimens with $1.5 \times 3 \times 15$ mm were cut from the central part.

Recrystallized SiC [2.85] exhibits a higher oxidation resistance than self-bonded SiC (Fig. 2.29). On the TG curve a mass gain of an AnnaNox CK ceramic specimen is detected only above $1300°$ C. The investigations under isothermal conditions (Fig. 2.30) demonstrated that the oxidation of recrystallized SiC started at temperatures close to $1000°$ C. A considerable mass gain as a result of the oxidation of specimens under isothermal conditions is caused by the fact that they possess an open porosity of up to 19% as opposed to dense self-bonded SiC. Therefore, their real reactive surface area is much larger than the geometric one.

However, at $1400°$ C the oxidation of self-bonded SiC is so rapid that the mass gain grows in time to a greater extent than for recrystallized ceramic specimens with a larger reactive surface area. At the same time the mass variation of an AnnaNox CK ceramic specimen (Fig. 2.29b) detected by a thermoanalyser at a relatively high heating rate is smaller than for self-bonded SiC over the whole investigated temperature range. In this case the oxidation of SiC is very severe at the initial stage of the process; after the formation of an oxide film it retards. But for recrystallized ceramics such rapid oxidation is not observed at the initial stage of heating; thus, the TG curve shows a very small mass variation of the specimens. AnnaNox CK and Crystar materials of similar composition and structure exhibit a similar oxidation behaviour. But for AnnaNox CK ceramics the mass variation after oxidation in the range of $1200°$–$1400°$ C is somewhat smaller (Fig. 2.30b,c). The appearance of the specimens of these materials does not change even after oxidation at $1500°$ C, and only a microscope can detect a very thin oxide film formed on the surface.

The rate constants calculated by equation (2.8) from the data of Fig. 2.30 are given in Table 2.8.

However for recrystallized ceramics the values of K_p are relative, since the calculations by (2.8) take into account the geometric surface area of the specimens, which did not coincide with the real reactive surface area. It is virtually impossible to estimate this value during calculations, as SiC oxidized to SiO_2 increases its volume by 107.4% (Table 2.1). The oxide phase fills pores in the

Table 2.8. Parabolic rate constants $K_p \times 10^4$, $mg^2/(cm^4 s)$ for SiC ceramics

T,° C	1000	1100	1200	1300	1400
Self-bonded SiC	–	0.02	0.5	4.3	10.4
AnnaNox CK	0.11	0.52	1.21	–	2.65
Crystar	0.18	0.59	2.22	–	2.92

specimen and therefore the real reactive surface area of the specimens changes continuously during oxidation, while (2.8) includes the value of their initial surface area. The values of K_p for self-bonded SiC and other similar materials are alike [2.79]. It should be noted that the kinetic curves of these and some others materials [2.71, 78] are described by (2.8) not exactly enough, especially during the first hours of oxidation. In [2.71] the equation $\Delta m = Kt^n$ (where n varies from 0.31 to 0.98 depending on oxidation temperature) was used to describe the oxidation kinetics of polycrystalline SiC in oxygen. But this equation is an empirical approximation. The curves plotted in the $\Delta m - \lg t$ coordinates possessed certain bends. In [2.86] an attempt was made to describe the obtained curves by Akimov's theory or the electrochemical theory by Frantsevich, which considers SiC as an anode, oxygen adsorbed on the surface of an oxide film as a gas cathode, and the oxide film as an electrolyte. It has been established that the experimental data are well described by the electrochemical oxidation theory.

The film formed on the surface of self-bonded SiC (Fig. 2.31) is amorphous (XRD analysis did not reveal any crystalline oxide phases in its composition). An X-ray microprobe analysis established that the calcium and aluminium contents in the surface layer increased abruptly after oxidation (Fig. 2.32) due to the diffusion of impurities with high oxygen affinity to the surface of the specimen during oxidation.

A petrographic analysis of the oxidation products of self-bonded SiC at 1450° C demonstrated the presence of several amorphous phases of different composition, with their refractive indices varying from 1.530 to 1.546. Apparently, these are calcium and magnesium aluminosilicates making up 55%–60% of the oxide layer. Another product of oxidation detected by petrographic analysis is amorphous silica with a refractive index of 1.455.

To elucidate comprehensively the mechanism of the oxidation processes of silicon carbide ceramics, the fracture surfaces of specimens after bending tests at different temperatures were studied. Since after fracture the specimens were cooled in air rapidly enough, one could detect on newly made fractures the changes which occurred during the first minutes of oxidation.

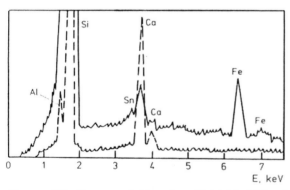

Fig. 2.32. EDAX spectra of the surface of initial (solid lines) and oxidized at 1350° C (dashed lines) specimens of self-bonded SiC

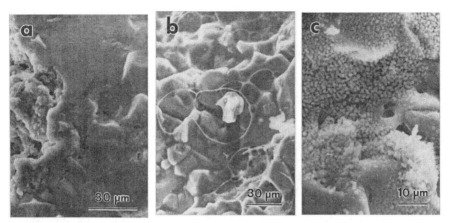

Fig. 2.33. Fracture surfaces of self-bonded SiC specimens tested at 1050° (a), 1250° (b) and 1350° C (c)

Figure 2.33a shows that one cannot find any traces of oxidation on the fracture surface of a self-bonded SiC specimen at 1050° C. At the same time the fracture of the specimen tested at 1250° C (Fig. 2.33b) is covered with a thin film of a blistered solidified liquid phase. A liquid aluminosilicate phase appears on the material due to active diffusion of impurities to its surface at this temperature [2.87].

On the fracture of the specimen tested at 1350° C (Fig. 2.33c) an oxide phase is actively formed, i.e. the SiC grains themselves are oxidized. The nuclei are not of regular crystalline shape, they are spherical bodies. The nuclei are not uniformly distributed over the whole surface of the specimen, but occupy only a certain part of it. It is characteristic that they are not always formed on the defects of structure, as it could be expected. Many grain boundaries and pores are not covered with such nuclei, while on outwardly smooth and defect-free surfaces the nucleation is very active.

A liquid phase formed above 1100° C gives evidence of the diffusion of impurities with a high affinity for oxygen to the surface. And the diffusion rate increases with temperature. At 1100°–1150° C the material starts to oxidize weakly, which is accompanied by the formation of calcium and magnesium aluminosilicates on its surface. This stage of the process corresponds to the area of the first exothermal peak on the DTA curve. However, because of the non-isothermal nonequilibrium conditions of the process, all the peaks in Fig. 2.29 shifted to higher temperatures. Thus, under isothermal conditions the second stage of more active oxidation starts above 1300° C, whereas on heating at a rate of 10° C/min this stage does not start below 1400° C. And the specimen surface is covered with a blistered film which is a mixture of amorphous silica and calcium and magnesium aluminosilicates (Fig. 2.31).

As opposed to self-bonded SiC, the specimens of recrystallized SiC are covered after oxidation with a very thin film of amorphous silica protecting their surface from further oxidation without deteriorating their quality.

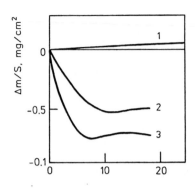

Fig. 2.34. Oxidation kinetics of self-bonded SiC in an argon/oxygen mixture at a gas flow rate of 38 l/h and 1150° (1); 1250° (2) and 1350° C (3) [2.89]

The isothermal studies on the oxidation kinetics of self-bonded SiC with lower contents of impurities and silicon at 1500° and 1650° C demonstrated [2.79] that at 1500° C the material exhibited a very high oxidation resistance. At 1650° C the oxidation rate increases, which is explained not only by a higher diffusion rate of oxygen and carbon oxides through the SiO_2 layer, but also by the poorer protective properties of this layer because of its discontinuity (blisters and pores) due to reaction (2.17) and the reaction $Si + SiO_2 = 2SiO$. Nevertheless, reactions (2.14, 15) prevail in this temperature range, since the specimens gain mass during oxidation. It should also be noted that at 1650° C pore-free self-bonded SiC is almost equal in its oxidation resistance to single crystal SiC [2.79].

It is also reported that self-bonded SiC of Nippon Tungsten (Japan) [2.88] and Ceranox CS600 and Ceranox CD100 grades of AnnaWerk (Germany) [2.89] exhibit a high oxidation resistance in air at temperatures up to 1350° C. At the same time, [2.89] cites abnormally low values of activation energy for the oxidation of SiC and silicon: 80–100 kJ/mole.

Both for silicon nitride and silicon carbide the passive–active oxidation transition occurs at low pressures and high temperatures. This process is accompanied by a mass loss of the specimens according to reaction (2.16) and the reaction $2Si + O_2 = 2SiO$. Here, more active oxidation of silicon results in the formation of pores in the areas of its inclusions. At high temperatures and high gas flow rates the mass loss becomes linear. At lower temperatures and lower gas flow rates mass gain is observed after several hours of oxidation (Fig. 2.34), it is caused by the formation of silica whiskers and fibers, which are 0.5–0.6 μm thick and up to 5 mm long, on the surface of the specimens [2.89].

2.3 Aluminium Nitride Ceramics

Aluminium nitride which is the most promising material after Si_3N_4 and SiC for manufacturing thermally stressed components of different high-temperature devices [2.90] and is used as an adittive in Si_3N_4 materials [2.72], has been studied far less. Thermodynamic calculations [2.9] have demonstrated that over the whole considered temperature range at atmospheric and low pressures the reactions leading to the formation of Al_2O_3 and nitrogen or its oxides are the most probable:

$$4\,AlN + 3\,O_2 = 2\,Al_2O_3 + 2\,N_2 \qquad (2.18)$$

$$2\,AlN + 2\,O_2 = Al_2O_3 + N_2O \qquad (2.19)$$

$$4\,AlN + 5\,O_2 = 2\,Al_2O_3 + 4\,NO \qquad (2.20)$$

$$4\,AlN + 7\,O_2 = 2\,Al_2O_3 + 4\,NO_2 \qquad (2.21)$$

Reaction (2.19) is characterized by the most negative change of Gibbs energy.

It should be noted that AlN, just as SiC, exists as several polytypes (2H, 165R and others [2.72]). The variation of a polytypic composition can also influence its oxidation resistance, though such data have not been found in the literature. According to [2.91] the oxidation rate of AlN is largely dependent on oxygen pressure and is proportional to $P_{O_2}^{1/4}$.

Sintered AlN with 8%–10% porosity starts oxidizing in oxygen at atmospheric pressure above 800° C and this process obeys the parabolic law (Fig. 2.35a). According to [2.92], at the initial stage the mass gain can be related linearly to the oxidation time.

The apparent activation energy of oxidation in the range of 900°–1100° C is $\sim 270\,kJ/mole$ [2.93]. As in the case of other hot-pressed and sintered AlN materials without additives [2.94, 95], the oxide layer reveals only α-Al_2O_3. Though on oxidation of AlN powders, γ- and δ-Al_2O_3 were found as intermediate oxidation products below 1000° C [2.92]. The IR spectroscopic analysis of the gaseous oxidation products detected NO and NO_2, apart from N_2, which suggests that reactions (2.20, 21) along with (2.18) are possible.

Dense hot-pressed AlN exhibits a higher oxidation resistance than sintered AlN [2.93]. In this case the kinetic curves of oxidation in air up to 1500° C are also described by the parabolic equation (Fig. 2.35b). At higher temperatures the relation approaches the linear one, because the oxide layer formed on the specimen surface is not dense but porous, with poorer protective properties. A dense α-Al_2O_3 layer formed at temperatures up to 1500° C cracked after cooling

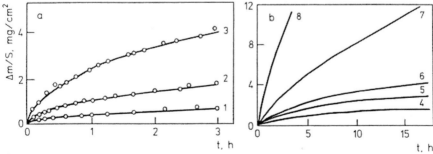

Fig. 2.35. Oxidation kinetics of sintered AlN in oxygen (a) and hot-pressed AlN in air (b) at 900°(1), 1000°(2), 1100°(3), 1300°(4), 1420°(5), 1490°(6), 1620°(7), and 1680° C (8)

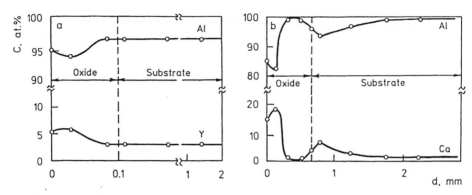

Fig. 2.36. Distribution of elements across the thickness of the specimen of hot-pressed AlN with Y_2O_3 after oxidation for 48 h at 1400° C (a) and with CaO after oxidation for 100 h at 1400° C (b) [2.97]

[2.94]. The oxidation process of AlN sintered at high pressures (1–5 GPa) is of a similar nature [2.96].

In addition to α-Al_2O_3 the oxide layer on additive-containing hot-pressed materials reveals $CaO \cdot 2Al_2O_3$, $CaO \cdot 6Al_2O_3$ when CaO is added, and $3Y_2O_3 \cdot 5Al_2O_3$ and $YAlO_3$ when Y_2O_3 is added [2.97]. And the content of yttrium and calcium in the surface layer exceeds their average concentrations in the material (Fig. 2.36), as in the case of oxidation of Si_3N_4 and SiC materials. According to the data provided in [2.98], titanium and niobium compounds added to hot-pressed AlN increase its oxidation resistance in air at temperatures below 1300° C due to the formation of a protective oxide layer consisting of titania, α-Al_2O_3 and aluminium niobates of indefinite composition.

In [2.99] the oxidation of sintered AlN with a porosity of < 3% was investigated during heating with a CO_2-laser (wavelength 10.6 μm, radiation power up to 800 W) in air. It is shown that heating up to 1300°–1400° C results in oxidation to a depth of 2–40 μm. At 1530° C an active exothermal reaction starts; it is accompanied by a self-heating of the specimen surface. Upon attaining the melting temperature of Al_2O_3, a liquid film formed on the surface retards oxidation and reduces heat release. If the radiation power is maintained constant, the surface temperature drops, the oxide crystallizes, cracks and spalls off. This is accompanied by a heating-up of the specimen surface. It is very important that for AlN, as opposed to Si_3N_4 and SiC, mass loss and active oxidation at low partial oxygen pressures are not observed. In this case the protective Al_2O_3 layer is also formed [2.100].

In conclusion we should note that already at 600° C the oxidation process accompanied by friction results in the formation of an α-Al_2O_3 layer up to 1 μm on the AlN surface [2.101].

2.4 Boron Carbide Ceramics

Boron carbide ceramics are rather promising because of their very high hardness and strength. But until quite recently they were mainly used at relatively low temperatures [2.72], therefore their oxidation has not been studied as thoroughly as that of the Si_3N_4 and SiC ceramics. There are data on the oxidation resistance of B_4C powders [2.102]. A number of papers devoted to the investigation of boron carbide materials produced by different methods have also been published [2.103–107].

In [2.9] the thermodynamic probability of different reactions proceeding in the system $B_4C - O_2$ was calculated at an oxygen pressure of 105 and $1.3 \cdot 10^{-3}$ kPa. It was shown that the oxidation of B_4C may result both in condensed (B_2O_3, B, C) and in several gaseous (CO, CO_2, B_2O_2, BO, BO_2) products. However, the most negative change of Gibbs energy is characteristic of the reaction

$$B_4C + 4\,O_2 = 2\,B_2O_3 + CO_2 \quad . \tag{2.22}$$

Such a reaction is completely confirmed by the results of experimental investigations of powders and sintered specimens [2.5].

Studies on hot-pressed B_4C containing iron, aluminium, magnesium, calcium, titanium, silicon and other impurities at a level of up to 4% and oxidized in air [2.103] demonstrated that upon non-isothermal heating of specimens the process started above 600° C. In the temperature range of 600°–1000° C on the DTA curve (Fig. 2.37) a broad exothermal peak is observed. A detectable mass gain, as the TG curve demonstrates, starts above 700° C. By the DTA data, at 1000° C the oxidation process becomes more active, and above 1200° C the material starts oxidizing considerably, with a high thermal effect. At temperatures above 1200° C the specimen losses mass due to the B_2O_3 vaporization (TG curve), with its rate increasing abruptly at this temperature [2.5].

The XRD analysis of oxidized specimens did not reveal any other solid oxidation products apart from B_2O_3. This corresponds to the thermodynamic calculations given in [2.72] and the experimental investigations [2.107] showing that during B_4C oxidation reaction (2.22) is probable.

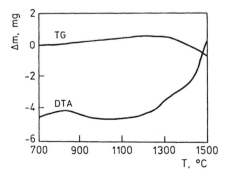

Fig. 2.37. Thermograms of oxidation of hot-pressed B_4C

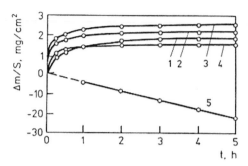

Fig. 2.38. Oxidation kinetics of hot-pressed B_4C at 800°(1); 900°(2); 1000°(3); 1100°(4) and 1250° C (5)

The kinetic curves obtained under isothermal conditions and presented in Fig. 2.38 are in agreement with thermal analysis data. Up to 1100° C the specimens gain mass with time, which obeys the logarithmic law (2.9). But the curves (Fig. 2.38) display two simultaneous processes, one of them increasing (B_4C oxidation) and the second one decreasing (B_2O_3 vaporization) the mass of specimens.

Due to the boron oxide vaporization at temperatures above 1200° C after the first 30 min of oxidation the mass loss of specimens (Fig. 2.38) becomes linear in time.

As a result of oxidation at temperatures below 1100° C a liquid boron oxide layer (T_m of amorphous B_2O_3 is 450° C [2.108]) uniformly covers the specimen surface. Upon heating of specimens to higher temperatures boron oxide after cooling solidifies as drops and beads in their lower part. Great differences in the thermal expansion coefficients of B_2O_3 and B_4C lead to cracking of the oxide layer after the cooling of the specimens (Fig. 2.39a). The oxidized specimens kept in air are covered with a white film due to the hydration of boron oxide (Fig. 2.39b).

The X-ray microprobe analysis of oxidized and initial specimens did not reveal higher contents of impurities in the surface layer after oxidation. Only XPS demonstrated a certain increase in the intensity of Si and Ca peaks after oxidation (Fig. 2.40). Under the conditions of active B_2O_3 vaporization the diffusion of impurities does apparently not exert great influence on the oxida-

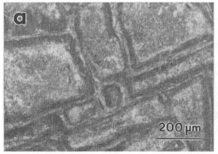
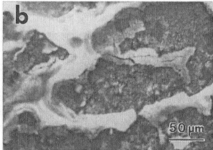

Fig. 2.39. Surface areas of hot-pressed B_4C specimens after oxidation at 800° C for 3 h (a) and after oxidation for the same time at 1000° C and further two-week exposure to air at room temperature (b)

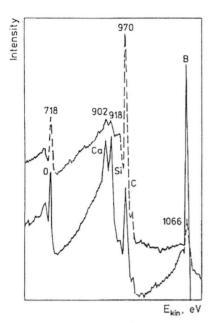

Fig. 2.40. XPS spectra obtained from the surface of an initial specimen (dashed lines) and the oxidized one at 1350° C (solid lines)

tion process. One may suppose that up to 1200° C the oxidation rate of B_4C is limited by the oxygen diffusion rate through the B_2O_3 layer. With the increase of temperature from 700 to 1200° C the B_2O_3 vaporization exerts a growing influence on the oxidation rate. Above 1200° C the rate of boron oxide vaporization becomes higher than the rate of its formation due to carbide oxidation, and subsequently the process is limited by the rate of chemical interaction of B_4C with oxygen of the air according to reaction (2.22).

In [2.107] it is assumed that at the first stage of oxidation (up to 1200° C) together with the diffusion of oxygen to the B_2O_3/B_4C interface, carbide components diffuse to the B_2O_3/air interface, with the diffusion rate of carbon being higher than that of boron. The variation in composition of the surface layer of B_4C leads to a change of its microhardness. Thus, the microhardness of the material after oxidation at 1100° C for 3 h and grinding of the oxide layer decreases from 36 to 24 GPa.

The B_2O_3 layer is removed from the surface of oxidized specimens by boiling in distilled water. Thus, the changes in the morphology of the surface layer due to oxidation can easily be detected. The oxidation of such ceramics causes the etching of grain boundaries and, thus, leads to a larger number of defects of their surface layer (Fig. 2.41).

As for other materials, a higher porosity of boron carbide specimens increases their reactive surface area and mass gain on oxidation (Fig. 2.42).

All the above refers to materials containing only relatively small amounts of impurities. However, boron carbide ceramics usually contain 7%–20% Al or Al_2O_3 to increase hardness and 6%–13% silicon to improve mechanical characteristics [2.72]. Aluminium- and silicon-doped materials often exhibit struc-

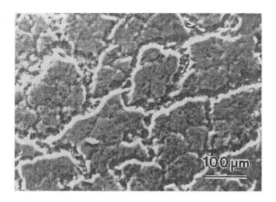

Fig. 2.41. Surface area of a hot-pressed B₄C specimen after oxidation at 1000° C and removal of oxide layer

tural nonuniformity. They possess zones of pure boron carbide, and silicon- and aluminium-doped areas [2.105]. Silicon and aluminium are present as SiC, B₄Si, AlB₁₀, AlB₁₂, Al₄SiC₄ and other compounds. The thermograms of oxidation of a material (Quartz-α-Silice, France) with a porosity of $\sim 6\%$ (2) and a material (Elektroschmelzwerk Kempten, Germany) with a porosity of $\sim 1\%$ (1) are presented in Fig. 2.43a,b respectively. Ceramics 2 revealed ~ 15 and ceramics 1 ~ 6 vol.% of aluminium and silicon compounds. These materials had different porosity but the same density of $2510\,\mathrm{kg/m^3}$ because of a different amount of additives.

The oxidation process of the more porous material 2 (Fig. 2.43a) with a higher amount of additives starts at 550° C and exhibits an intensive exothermal peak on the DTA curve, though a detectable mass gain due to the oxidation of B₄C according to reaction (2.22) occurs only from 600° C. A protective oxide layer formed on the specimen surfaces at 640° C abruptly retards the rate of

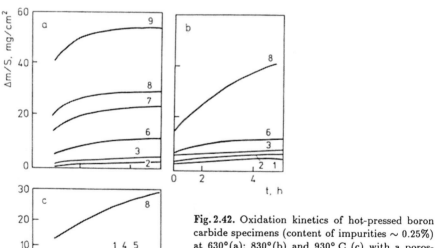

Fig. 2.42. Oxidation kinetics of hot-pressed boron carbide specimens (content of impurities $\sim 0.25\%$) at 630°(a); 830°(b) and 930° C (c) with a porosity of 0%(1); 1%(2); 6%(3); 8%(4); 12%(5); 19%(6); 33%(7); 37%(8) and 48%(9) [2.106]

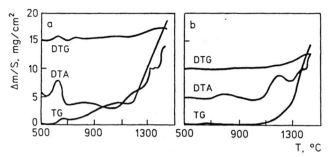

Fig. 2.43. Thermograms of oxidation of materials 2(a) and 1(b)

the process. Only at $\sim 775°$ C is a mass gain observed again, probably due to the beginning of oxidation of silicon carbide as a component of the material.

However, in this case instead of an exothermal peak, a broad endothermal peak determined by the beginning of boron oxide vaporization appears on the DTA curve above $900°$ C. In the range of $1030°–1190°$ C on the TG curve a plateau is seen again, and only at $1200°$ C the process becomes much more active, as it was also noted in [2.103]. But, in contrast to the data presented in Fig. 2.37, for the additive-free material mass gain instead of its loss is observed, i.e. the rate of boron oxide formation is much higher than the rate of its vaporization. In the range of $1200°–1400°$ C on the DTA curve different thermal effects are observed. They give evidence of several simultaneous exothermal (oxidation of B_4C, SiC and other components of the material) and endothermal (vaporization of B_2O_3) processes.

For the denser material 1, which contains a smaller amount of additives, a weak exothermal peak corresponding to the beginning of oxidation was noted above $700°$ C (Fig. 2.43b). In this case the mass variation of specimens is not detectable. A certain mass gain is detected only above $800°$ C, and rather intensive oxidation starts at $1100°$ C, with the exothermal peak having its maximum at $1200°$ C, just as for material 2. At higher temperatures the DTA and TG curves for both materials are similar.

The somewhat lower oxidation resistance of material 2 at the initial stage of the process may be due to its higher porosity, which increases the real reactive surface area of the specimens.

Electron microscopy investigations of specimen surfaces of both materials heated to $1430°$ C demonstrated that the oxidized silicon- and aluminium-doped areas were covered with a glassy layer (Fig. 2.44a) containing boron, oxygen, silicon and aluminium. Thus, the oxide layer is borosilicate glass. The dissolution of silica and alumina in boron oxide abruptly retards the rate of its vaporization. At the same time on the areas of pure boron carbide B_2O_3 vaporizes actively and in these places the craters are formed (Fig. 2.44b,c). A similar picture was observed earlier for boron nitride oxidation [2.58]. The majority of craters on the specimens of both materials had knobs in the central part forming a rough surfac. On its top, apart from boron and oxygen, molybdenum and tungsten were revealed. These had been present in the initial materials as

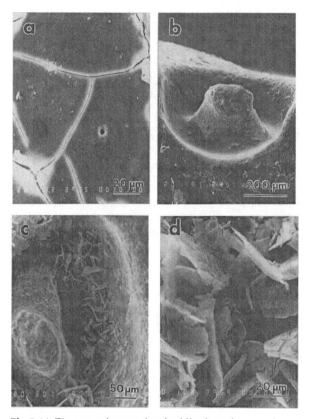

Fig. 2.44. Electron micrographs of oxidized specimen surfaces for materials 1 (a,b) and 2 (c,d)

small amounts of impurities. The formation of knobs in the centre of the craters is apparently caused by the transport of oxides of impurity elements through the gas phase. A small amount of iron was also found in the composition of the oxide layer.

The specimens of material 1 with smaller amounts of silicon and aluminium showed much more craters and less glassy phase beads after oxidation. The XRD analysis revealed B_2O_3 and H_3BO_3 in the surface layer of oxidized specimens. But H_3BO_3 is not formed during the oxidation of materials, as it was erroneously supposed in [2.109], but during the exposure of specimens to air at room temperature for several days due to the hydration of boron oxide according to the reactions $B_2O_3 + H_2O = 2\,HBO_2$ and $HBO_2 + H_2O = H_3BO_3$.

Therefore, the lamellar crystals of boron hydroxide were formed in the craters (Fig. 2.44c) and on several pure B_2O_3-enriched surface areas (Fig. 2.44d).

It is interesting to note that we investigated the specimens of ceramics oxidized at $1400°$ C by XRD and AES revealed a graphite layer of up to $30\,\mu$m thickness under the B_2O_3 film [2.105]. If this process was accompanied by friction, the graphitized layer was formed at much lower temperatures.

The addition of silicon and aluminium to boron carbide facilitates the formation of borosilicate glass on oxidation and decreases the rate of boron oxide vaporization, thus increasing the oxidation resistance of the material.

However, to form a continuous protective layer on the specimen surface, it is necessary to provide a uniform distribution of additives in the material excluding the existence of pure boron carbide areas.

2.5 Boron Nitride Ceramics

Materials based on different crystalline modifications of boron nitride find wide application in manufacturing refractory furniture, cutting tools [2.110], components used in MHD generators and space equipment [2.72]. Boron nitride also appears as one of the components in Si_3N_4 [2.56] and AlN materials.

Thermodynamic calculations [2.9] show that the oxidation of boron nitride probably proceeds by the reactions

$$4\,BN + 3\,O_2 = 2\,B_2O_3 + 2\,N_2 \tag{2.23}$$

$$2\,BN + 2\,O_2 = B_2O_3 + N_2O$$

$$4\,BN + 5\,O_2 = 2\,B_2O_3 + 4\,NO$$

$$4\,BN + 7\,O_2 = 2\,B_2O_3 + 4\,NO_2$$

Over the whole temperature range from 0° to 2000° C at atmospheric and low pressures reaction (2.23) is the most probable. Under these conditions the reactions of formation of lower boron oxides BO, BO_2 or boron are characterized by a positive change of Gibbs energy. The oxidation of boron nitride as well as boron carbide is distinguished by an oxide layer remaining in a liquid state and vaporizing considerably over the whole temperaturre range.

However, the study of the mechanisms of oxidation of boron nitride materials is complicated by the existence of several BN modifications with considerable differences in structure and properties. The oxidation of a stable (under normal conditions) hexagonal α-modification of BN with a graphite lattice has been studied most thoroughly [2.58].

The oxidation of BN powders starts above 700° C and results in the formation of B_2O_3. The activation energy of oxidation of a disordered (turbostratic) BN powder is 309 ± 21 kJ/mole. But depending on the structure ordering (Fig. 2.45), the behaviour of α-BN materials changes considerably on heating in air. With growing ordering of a crystalline lattice, the temperature of a maximum oxidation rate also grows. Thus, an exothermal peak on the DTA curve is observed for mesographite BN at 1080° C and for graphite-like BN at 1180° C. A well-pronounced maximum for pyrolytic turbostratic BN has not been revealed. Thus, even taking into account a high porosity of graphite-like BN (Table 2.9), one should come to the conclusion of a predominant influence of structural defects in the crystalline lattice on the oxidation of α-BN. An ordered arrange-

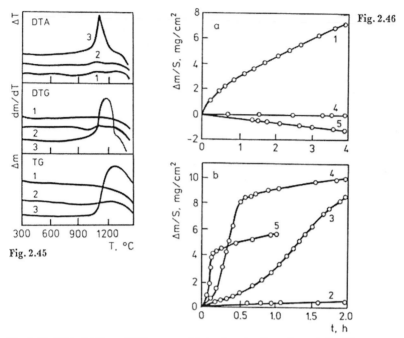

Fig. 2.46

Fig. 2.45. DTA, DTG and TG curves obtained on heating of pyrolytic (1), mesographite (2) and graphite-like (3) α-BN in air

Fig. 2.46. Oxidation kinetics in oxygen at atmospheric pressure for pyrolytic (a) and graphite-like (b) boron nitride at 900°(1); 950°(2); 1000°(3); 1100°(4) and 1200° C (5)

ment of hexagonal layers in the α-BN lattice strengthens the bonds between layers and within the layer, which, in turn, increases the energy of a B–N bond break. However, the strength of a chemical bond has an influence only at the initial stage of oxidation.

As was shown earlier, upon long exposure under isothermal conditions everything is determined by the composition and real microstructure of the material and the oxide layer formed. For pyrolytic BN a mass loss due to the B_2O_3 vaporization could be observed already at 900° C, increasing with temperature [2.111], while for porous specimens of graphite-like BN there was a mass gain up to 1200° C (Fig. 2.46). This is apparently due to inner oxidation of porous specimens. Here the whole bulk of the specimen is oxidized due to rather active diffusion of oxygen through the oxide layer with a very low viscosity; B_2O_3 vaporized only from the surface up to its boiling point (1500° C).

When the oxidation products vaporize from the surface of a dense specimen, the kinetic curves, as in the case of B_4C, are linear, and the rate of the process is limited by the rate of interaction of the material with air. The kinetic curves of oxidation of graphite-like BN presented in Fig. 2.46 cannot be described by any of the known equations. The bends on the curves are probably due to the retardation of oxidation after the filling of pores with melted B_2O_3 in the

Table 2.9. Characterization of boron nitride specimens

Characteristics	α-BN			Wurtzite-like BN
	pyrolitic turbostratic	meso-graphite	graphite-like*	
Sizes, mm	10 × 15 × 1	3 × 5 × 2	10 × 10 × 3	$d = 4\ h = 7$
Content**, %				
O_2	–	0.2	0.2	0.3
N_2	53.45	55.90	55.90	–
Density, kg/m³	1850	2180	1460	3400
Porosity, %	20	5	35	0

* The specific surface area of graphite-like α-BN is 0.117 m²/g
** By spectral analysis data the content of impurities varied, %: Mg ~ $10^{-3} - 10^{-2}$; Al ~ $10^{-3} - 10^{-2}$; Si ~ $10^{-3} - 10^{-1}$; Ti ~ $10^{-3} - 10^{-2}$; Fe ~ 10^{-2}; Ca ~ $10^{-2} - 10^{-1}$

subsurface layer (from the surface layer B_2O_3 continuously vaporizes at such temperatures).

The kinetic curves of oxidation of pyrolytic BN in oxygen and in air are described by the paralinear equation (not the mass variation of specimens but the thickness of the oxide layer x was recorded) [2.112]. The paralinear kinetics describes the processes in which the reaction stage linearly related to time $x = K_l t$ (K_l is the linear constant) is superimposed on the parabolic diffusion stage. In this case the relation of the thickness of an oxide film x to the oxidation time t takes the form

$$ t = \frac{K_p}{K_l^2} \ln \left(\frac{K_p}{K_p - K_l x} \right) - \frac{x}{K_l} \quad , $$

where K_p is the parabolic constant.

It should also be noted that the oxidation of pyrolitic BN revealed etching pits on basal a-planes in the areas of the highest density of screw dislocations, which coincided roughly with the directions of the axes of nitride growth cones [2.58].

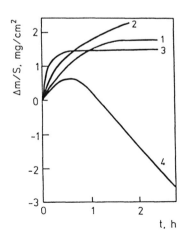

Fig. 2.47. Oxidation kinetics of Hexanite–R at 900° (1); 1000° (2); 1100° (3) and 1200° C (4)

The studies on Hexanite–R ceramics based on wurtzite-like boron nitride [2.113] demonstrated that the run of the kinetic curves of its oxidation (Fig. 2.47) was very similar to that of the kinetic curves for hot-pressed B_4C (Fig. 2.38). Rather high oxidation resistance is also exhibited by a cubic sphalerite-like boron nitride-based material (Elbor–R) and materials that are a mixture of wurtzite- and sphalerite-like BN [2.114].

Although materials based on the high-pressure phases possess very high hardness and strengh, they are unstable at high temperatures, which imposes restrictions on their applications. At present these materials are mainly used for manufacturing cutting tools. When they are heated in air, along with oxidation, the wurtzite- and sphalerite-like modifications are transformed into the graphite-like one [2.110].

2.6 Ceramic Matrix Composites

The majority of ceramic materials described in the preceding sections of this chapter are single-phase or contain the secondary grain boundary phase. At present a new generation of ceramic materials, composite ceramics, is effectively being developed [2.115]. Such a material is produced by adding to one of the above ceramics 10%–50% of other refractories as whiskers, fibers or particles greatly differing in their properties from the matrix. This may increase the strength and fracture toughness but it can also exert great influence on the corrosion resistance of these ceramics. Above we discussed several similar materials, e.g. NKKKM ceramics.

From the point of view of the behaviour in oxidizing environments the ceramic composites may be divided into four large groups:

1) oxide matrix–nonoxide reinforcer (e.g. $Al_2O_3 - SiC$);
2) nonoxide matrix–nonoxide reinforcer with a lower oxidation resistance ($Si_3N_4 - TiN$);
3) nonoxide matrix–nonoxide reinforcer with equal or higher oxidation resistance ($Si_3N_4 - SiC$);
4) nonoxide matrix–oxide reinforcer ($Si_3N_4 - ZrO_2$).

As it was shown in Sect. 2.1 and in [2.116], the ceramics of groups 3 and 4 are not critically different in their oxidation behaviour from the matrix material. The problems in the case of $Si_3N_4 - ZrO_2$ ceramics were brought about by the formation of zirconium oxynitride [2.49], but not by the presence of transformation-toughening ZrO_2 particles in the silicon nitride matrix. At the same time the addition of reinforcing phases to the ceramics of groups 1 and 2 may impose restrictions on high-temperature applications because of a lower oxidation resistance. Therefore we dwell on them in more detail.

Table 2.10. Characterization of materials

No	Additives, %		Preparation method	Phase composition	Density, kg m^{-3} (% of theoretical shown in brackets)	Strength at 20° C, MPa	Ref.
	Nonoxide refractory compound	Sintering aid					
1	0–50 TiN	5 Y$_2$O$_3$+ 2 Al$_2$O$_3$	Hot pressing	β-Si$_3$N$_4$, TiN, Si$_2$N$_2$O, WSi$_2$	(99)	700–1100	[2.117]
2	20 ZrN	–	Pressureless sintering	β-Si$_3$N$_4$, ZrN	2330 (64)	not measured	[2.118]
3	30 TiC$_{0.5}$N$_{0.5}$	–	Pressureless sintering	β-Si$_3$N$_4$, TiCN	2550 (67)	not measured	[2.118]
4	10 TiC$_{0.5}$N$_{0.5}$	10 Al$_2$O$_3$	Pressureless sintering	β-Si$_3$N$_4$, x$_1$ (J-phase), x$_2$ (15R), TiCN	3550 (> 98)	not measured	[2.118]
5	42 TiN	15 Al$_2$O$_3$+ 8 AlN	Hot pressing	β'-Si$_3$N$_4$, x$_1$, WSi$_2$, TiN	3850 (> 99)	600	[2.119]

2.6.1 Nonoxide Matrix Composites

Silicon Nitride Composites. We investigated quite a number of particulate composites (Table 2.10) considerably differing in their composition and properties to elucidate the effect of refractory additives on the oxidation resistance of silicon nitride ceramics.

Pressureless-sintered silicon nitride ceramics, containing ZrN and TiC$_{0.5}$N$_{0.5}$, are oxidized more actively during heating in air. The TG curve of Fig. 2.48 shows that the mass gain of material 2 starts at 370° C and that up to 700° C oxidation is very rapid, with a peak on the DTG curve and an exothermal peak on the DTA curve. The XRD analysis of the specimens oxidized at 600° C under isothermal conditions revealed cubic ZrO$_2$. Under these conditions Si$_3$N$_4$ was not oxidized. Thus, within the above-mentioned temperature range the reaction proceeds as follows:

$$2\,ZrN + 2\,O_2 \rightarrow 2\,ZrO_2 + N_2 \tag{2.24}$$

As calculations based on (2.24) demonstrate, more than 80% of the ZrN is oxidized on heating up to 700° C under non-isothermal conditions. Probably, only particles isolated from the ambient atmosphere by Si$_3$N$_4$ grains remain unoxidized. Cracking of the specimens (shown in Fig. 2.49) due to strong internal stresses resulting from the ZrN $\rightarrow$ ZrO$_2$ transformation and the phase transitions in ZrO$_2$ also promotes almost complete ZrN oxidation. Since the

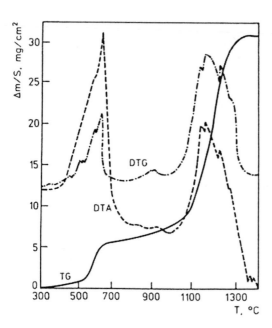

Fig. 2.48. Thermograms of oxidation of material 2, a silicon nitride ceramic containing 20% ZrN (heating rate 15° C/min)

specimens crack during isothermal exposure or heating, but not cooling, the process is due more to the $ZrN \rightarrow ZrO_2$ transformation than the phase transitions in ZrO_2. The cracking evidently activated oxidation and caused the appearence of sharp peaks with a maximum at 650° C on the DTA and DTG curves, after the rate of the process started to decrease. Severe oxidation of the material in the temperature range investigated indicates poor protective properties of the oxide layer, despite an increase in volume due to the $ZrN \rightarrow ZrO_2$ transformation.

When the specimen is heated above 700° C, the rate of mass gain initially decreases (up to 800° C), since oxidation of ZrN is virtually complete, and then increases due to Si_3N_4 oxidation, which proceeds in two stages [2.52].

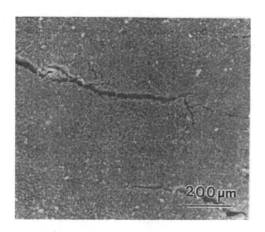

Fig. 2.49. Electron micrograph of the specimen surface of material 2, a silicon nitride ceramic containing 20% ZrN after heating up to 500° C

During the first stage, within the temperature range of 760°–1000° C, formation of silicon oxynitride dominates. It proceeds by the reaction

$$4\,Si_3N_4 + 3\,O_2 \rightarrow 6\,Si_2N_2O + 2\,N_2 \quad . \tag{2.25}$$

Very low peaks appear on the DTA and DTG curves (Fig. 2.48).

During the second stage Si_3N_4 and Si_2N_2O are oxidized by reaction (2.1) and the following reaction, respectively:

$$2\,Si_2N_2O + 3\,O_2 \rightarrow 4\,SiO_2 + 2\,N_2 \quad . \tag{2.26}$$

Sharp peaks on the DTA and DTG curves indicate this. Pronounced oxidation of Si_3N_4 is associated with the oxidation of ZrN without the formation of a protective oxide layer. In fact, cracking of the specimen occurs. Therefore the section of the thermogram under consideration resembles the curves for Si_3N_4 powders [2.52] more than those for porous silicon nitride. The complex shapes of the DTA and DTG curves show that several processes occur simultaneously during oxidation of the material under study at 1100°–1300° C. Three to four peaks overlap in this narrow temperature range, corresponding to reactions (2.1) and (2.26) as well as to the reaction

$$ZrO_2 + SiO_2 = ZrSiO_4 \quad . \tag{2.27}$$

Zircon was detected in the specimens heated up to 1400° C under non-isothermal conditions and in the specimens heated to 1100° C under isothermal conditions for 4 h. Although zirconia was found at 1100° C, after heating up to 1400° C it reacted completely with excess silica. In the surface layer of the specimens heated up to 1400° C α-cristobalite and a glassy phase were also present. Probably, rapid Si_3N_4 oxidation provides access for oxygen to unoxidized ZrN, and one of the peaks in this temperature range corresponds to the activation of reaction (2.24). During heating up to 1300° C the process is sharply retarded again; however, this is not due to complete nitride oxidation ($\sim$ 40%Si_3N_4 is oxidized) but to the formation of a protective silica- and zircon-containing layer. Oxidation of Si_3N_4 in this material does not cause further cracking of the specimens, but results in the partial healing of previously formed pores and cracks because the oxide layer has a much greater volume than oxidized Si_3N_4.

Oxidation of materials containing titanium carbonitride starts at about 380° C (Fig. 2.50), but an active mass gain due to the formation of TiO_2 as rutile takes place only above 600° C:

$$4\,TiC_{0.5}N_{0.5} + 5\,O_2 = 4\,TiO_2 + 2\,CO + N_2 \quad . \tag{2.28}$$

At temperatures up to 900° C lower titanium oxides can be formed at the rutile/carbonitride interface. Some data suggest the formation of an oxycarbonitride of the formula $TiC_xO_yN_z$.

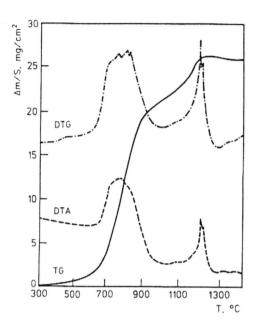

Fig. 2.50. Thermograms of oxidation of material 3, a silicon nitride ceramic containing 30% $TiC_{0.5}N_{0.5}$ (heating rate 15° C/min)

On the DTA and DTG curves of Fig. 2.50 complex peaks appear in this range and the $TiC_{0.5}N_{0.5}$ diffraction lines shift. A similar process also occurs during oxidation of the material containing ZrN.

Intensive titanium carbonitride oxidation is accompanied by the appearance of peaks on the DTA and DTG curves; it occurs over a wider temperature range than that of zirconium nitride. The process slows down only at 900° C when about 50% of the compound has been oxidized. Cracking of the specimens containing titanium carbonitride was not observed, although small cracks were detected on a few specimens oxidized under isothermal conditions at 600 and 800° C. $TiC_{0.5}N_{0.5}$ is therefore oxidized more slowly than ZrN, and the corresponding peaks on the DTA and DTG curves overlap the region of the first stage of Si_3N_4 oxidation which occurs by reaction (2.25).

A second exothermal peak on the DTA curve appears in the temperature range 1100°–1250° C, with a maximum at 1200° C. This peak coresponds to the oxidation of silicon nitride according to reactions (2.1) and (2.6), but – unlike materials containing ZrN – oxidation of the materials containing $TiC_{0.5}N_{0.5}$ is accompanied only by slight oxidation of Si_3N_4. This is due to the absence of cracks on the specimens after the first stage of oxidation and the protective effect of the oxide layer consisting of TiO_2 and SiO_2. Thus, oxidation of silicon nitride occurred mainly in the surface layer, and the mass gain of the specimens ceased at 1250° C. As in the previous cases, oxidation of Si_3N_4 led to the formation of α-cristobalite and amorphous silica.

For all the above materials oxidation resulted in a comparatively high mass gain, which can be explained by their porosity ($\sim$ 30%). Therefore oxidation proceeds over all the bulk, and the real reactive surface area exceeds the geometric surface of the specimen.

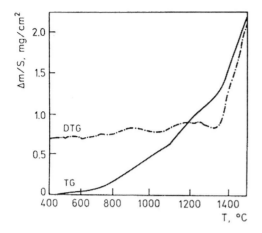

Fig. 2.51. Thermograms of oxidation of material 5, a silicon nitride ceramic containing 42% TiN, 15% Al_2O_3 and 8% AlN (heating rate 15° C/min)

Silicon nitride cannot be sintered to a high density without oxide sintering aids. The addition of 10% Al_2O_3 to the $Si_3N_4 - TiC_{0.5}N_{0.5}$ system made it possible to sinter without pressure materials with a density exceeding 95% of the theoretical one (Table 2.10). The specimens of these materials had no open porosity which determined their high oxidation resistance in air up to 1400° C. At 500°–800° C the mass gain over 4 h was approximately two orders lower than for the same material without Al_2O_3. The oxide layer formed on the specimen surface at 1400° C includes mullite, rutile and cristobalite. $TiC_{0.5}N_{0.5}$ inclusions on the specimen surface were oxidized with the formation of rutile islands up to 30 μm in size.

A material similar to the one described above but containing TiN instead of $TiC_{0.5}N_{0.5}$ and AlN as well as Al_2O_3, which ensures the formation of a sialon matrix consisting of β'-Si_3N_4 also exhibits a rather high oxidation resistance up to 1300° C (Fig. 2.51). The high TiN content, which has a lower oxidation resistance, determined a higher mass gain under non-isothermal heating compared with material 4. The oxidation mechanisms of materials 3, 4 and 5 are similar, since during oxidation titanium nitride and carbonitride behave similarly. Similar to Fig. 2.50, the DTG curve in Fig. 2.51 shows two peaks over the temperature range 600°–1300° C corresponding to the oxidation of TiN and Si_3N_4. However, the first peak shifts to higher temperatures and both peaks are rather weak and smeared, since oxidation occurs not in the bulk, but only on the specimen surface. The presence of rutile in the surface layer was detected above 700° C. At lower temperatures only a solid solution of TiN_xO_y is formed. As a result, TiN diffraction maxima shift to wider angles and the microhardness of the surface layer increases (oxynitride exhibits a higher hardness than nitride) [2.9]. Thus, after exposure at 600° C for 1 h the microhardness of the specimens increased from 16 to 18 GPa (Vickers pyramid, 2N-load); at 800° C oxynitride is formed in the subsurface layer and a rutile layer appears on the surface. Under these conditions oxidation of a sialon matrix is very weak. Therefore, after oxidation for 1 h cristobalite and mullite, the typical products of sialon oxidation [2.62], were not found in the surface layer of the specimens over the temperature range investigated. At the same time aluminium, having a high affinity for oxygen, diffuses from the interior to the surface, which is typical of sialons [2.62]. Above 1000° C Al_2O_3 interacts with TiO_2 to form aluminium titanate:

$$Al_2O_3 + TiO_2 \rightarrow Al_2TiO_5 \qquad (2.29)$$

A basic difference in the behaviour of material 5 on the one hand and materials 3 and 4 on the other hand is observed above 1300° C. A protective SiO_2-based layer is formed on materials 3 and 4 (due to active inner oxidation of Si_3N_4 in 3 and a low content of $TiC_{0.5}N_{0.5}$ in 4), whereas such a layer is not formed on material 5, and thus TiN oxidation occurs. The considerable increase in volume caused by the TiN $\rightarrow$ TiO_2 transformation leads to the formation of a continuous TiO_2 layer after heating up to 1500° C instead of islands or inclusions in the SiO_2 layer, as observed for materials 3 and 4. Moreover, the screening effect of this layer is so great that on the diffraction pattern of the surface of the specimen heated up to 1500° C phases other than rutile are not found. Such severe TiN oxidation at high temperatures is probably due to the discontinuity of the oxide layer because of strong internal stresses. A continuous skeleton formed by TiN grains in material 5 allows TiN to be oxidized in the interior.

From the results of thermal analysis, one can determine the type of chemical reactions occurring during oxidation and the temperature ranges at which these reactions occur. The kinetic parameters of oxidation can be calculated quite accurately from any curve presented in the thermograms. To illustrate this, we calculated the activation energies corresponding to the TiN and Si_3N_4 oxidation stages for material 5. The thermograms were recorded at three different heating rates (15, 7.5 and 3.75° C/min) [2.120]. The activation energy E was calculated from the plot of log (temperature growth rate) vs inverse temperature of the peak on the DTG curve (Fig. 2.52) using the equation derived by T. Ozawa

$$E = 2.19 \, Rd \log \beta / d(1/T_p) \quad , \qquad (2.30)$$

where R is the universal gas constant, β is the heating rate and T_p is the peak temperature [2.121]. Equation (2.30) is derived for DTA, but the coincidence of peaks on the DTA and DTG curves allows to use the DTG data if DTA curves were not recorded or maxima are not so pronounced because of weak thermal effects.

The activation energy of TiN oxidation in material 5 (227 kJ/mole) is slightly higher than that of TiN single crystals (195 kJ/mole) [2.116], TiN powder (173 kJ/mole) and hot-pressed $Si_3N_4 - TiN$ ceramics (160 kJ/mole) [2.122]. It confirms that the oxygen diffusion through the TiO_2 film is a limiting factor at the first stage of oxidation.

It is more complicated to interpret the result obtained for the second peak in Fig. 2.51. This peak corresponds to Si_3N_4 oxidation, but in the temperature range of interest a further TiN oxidation also proceeds. As shown previously, the mechanism of Si_3N_4 oxidation in this material differs considerably from

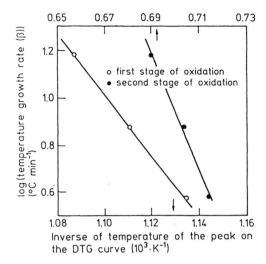

Fig. 2.52. The logarithm of heating rate (β) as a function of peak temperature for the first and the second stages of oxidation of material 5, a silicon nitride ceramic containing 42% TiN, 15% Al$_2$O$_3$ and 8% AlN

that of pure silicon nitride oxidation, since the SiO$_2$ layer is not formed on its surface, oxygen mainly combines with TiN and aluminium diffuses to the surface etc. Therefore, the value of $E = 455\,\mathrm{kJ/mole}$ is much higher than that calculated for oxygen diffusion into SiO$_2$ and the oxidation of Si$_3$N$_4$ powder, but they are within the experimental values obtained for different hot-pressed silicon nitride ceramics (Table 2.5).

The above results demonstrated the effect of composition and porosity of ceramic matrix composites on the mechanism and kinetics of their oxidation. To investigate the influence of the amount and fineness of a TiN additive on the oxidation resistance, specimens containing TiN of different fineness (material 1 in Table 2.10) were prepared. Industrial TiN powder was classified by sedimentation into average particle sizes of $d_{50} = 0.83, 2.43$ and $36.8\,\mu\mathrm{m}$. Silicon nitride with α-phase content $> 90\%$ was ball-milled with 5 % Y$_2$O$_3$ and 2 % Al$_2$O$_3$, $d_{50} = 1\,\mu\mathrm{m}$. This composition was mixed with TiN in a planetary mill for a very short time, to avoid grain size variations. One mixture was ball-milled to produce a uniform batch with minimum TiN particle sizes. The experiments showed that the oxidation resistance of ceramics decreased in comparison with pure HPSN when the TiN content was above 20% (Fig. 2.53). At the same time the oxidation resistance is strongly dependent on TiN grain sizes (Fig. 2.53) and abruptly increases with the decrease of TiN particle sizes.

Thus, ball-milled ceramics with the finest and uniformly distributed TiN particles possess high oxidation resistance at a TiN content of up to 30%. Up to 1000° C the oxide layer revealed rutile, at higher temperatures α-cristobalite, yttrialite, mullite and Y$_2$Ti$_2$O$_7$ could also be found. The ceramic specimens with a TiN content $> 30\%$, just as ceramics 5, were covered by a continuous rutile layer on heating up to 1500° C or exposure for 100 h at 1100° C or more (Fig. 2.54). This TiO$_2$ film is separated from the ceramic surface with a porous layer (Fig. 2.54a) which can delaminate easily enough. This will obviously have

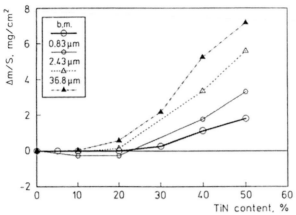

Fig. 2.53. Mass gain for 100 h at 1130° C as a function of TiN content with different average particle sizes (b.m. – ball-milled)

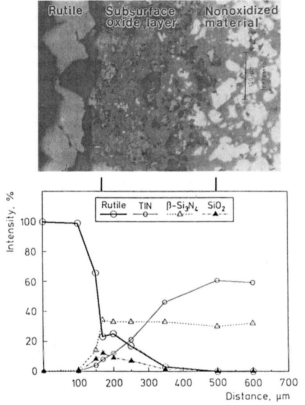

Fig. 2.54. Cross-section of an oxidized HPSN specimen with 50 mass% TiN (a) and semiquantitative XRD data (b)

an adverse effect on the performance of ceramics under real operating conditions. We should note that our results are in good agreement with the data of [2.122] obtained in the investigation of ceramics of a similar composition.

In [2.119, 120] the studies on oxidation of ceramics 5 during radiant heating up to 2100° C demonstrated that short-time heating above 1700° C resulted in the dissociation of Si_3N_4 and the interaction of silicon with TiN to form titanium silicide. This leads to the degradation of the surface layer of the material.

An analysis of the results shows that the behaviour of the silicon nitride-refractory compound system does not contradict the relationships obtained for the oxidation of individual components. Therefore SiC, having the same or even higher oxidation resistance, causes neither a decrease in oxidation resistance nor a noticeable effect on the composition of oxidation products when added to Si_3N_4. The addition of compounds that are easily oxidized in air results in lower oxidation resistance and can considerably influence both the oxidation mechanisms of the material and the composition and protective properties of the oxide layer. The presence of isolated inclusions of the above refractory compounds in the silicon nitride matrix does not change the mass gain during oxidation or increases it slightly, in comparison with a material without refractory compounds. For materials with such a structure the maximum operating temperature in the oxidizing environment is determined by the properties of the silicon nitride matrix. For materials containing a continuous skeleton of refractory compounds, this is the maximum temperature for the component with the least oxidation resistance.

Other Nonoxide Composites. In the literature there are data on the investigation of SiC-based composites [2.123]. Hot-pressed ceramics with 15 vol. % TiB_2 possessed higher strength (by 28%) and fracture toughness (by 45%). The ceramics retained the high oxidation resistance inherent in SiC up to 1200° C. At 1400° C a drastic oxidation started.

Sintered materials of the system $C-SiC-B_4C$ can also retain high oxidation resistance if they contain a proper amount of SiC and B_4C, since their ratio determines the content of SiO_2 and B_2O_3 in the oxide layer and its protective properties. It is reported in [2.124] that the oxidation resistance in an air stream at 1200° C of ceramics with an optimum composition is comparable to that of the silicon nitride ones.

Refractory compounds added to ceramics with low oxidation resistance, e.g. B_4C, often contribute to higher performance at high temperatures. Therefore, materials of the system $B_4C - B_4Si$ (Fig. 2.55) and materials with borides of transition metals [2.72] exhibit a higher oxidation resistance than pure B_4C.

In conclusion we should note that BN-based materials acquire a higher oxidation resistance if 5%–30% SiO_2 are added, just as in the case of B_4C [2.125].

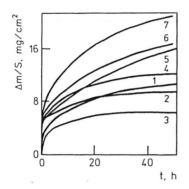

Fig. 2.55. Oxidation kinetics at 1000° C for materials of the system $B_4C - B_4Si$ with 100%(1); 90%(2); 70%(3); 50%(4); 30%(5); 10%(6) and 0% B_4Si (7) [2.72]

2.6.2 Oxide Matrix Composites

Most of the data published in the literature refer to Al_2O_3-based ceramics [2.126–129], but there are also publications devoted to ceramics of the system mullite–SiC [2.130] and composites of a more complex composition [2.131–134].

In [2.128] an attempt is made to calculate the rate constants of oxidation of Al_2O_3 – SiC and mullite–SiC composites using the data provided in the literature on the grain boundary and bulk diffusion. It was assumed that a SiC content of up to 24.4 vol. % in Al_2O_3 results in an oxide layer containing Al_2O_3 and mullite; when its concentration was higher the oxide layer consisted of silica and mullite. The calculated value of the activation energy for the Al_2O_3–20 vol. % SiC composite at 1400° C was 330–465 kJ/mole, which does not differ so much from the value of 502 kJ/mole obtained experimentally [2.130]. The authors of [2.130] also investigated other materials and established that the oxidation behaviour of the composites studied can be divided into the two main categories of either protective or non-protective scale formation. Non-protective scales result from the formation of a rigid, crystalline oxidation product whose volume exceeds that of the bulk material by at least 15%. Even if the scale does not spall off, the coating may be disrupted by strong internal stresses, thereby allowing oxygen to effuse rapidly to the interface. The disrupted coating also permits easy egress of any gaseous reaction products. In such cases (alumina-TiC and alumina-MoSi$_2$) the growth of the reaction product is linear with time and the rate controlling step is the reaction at the interface.

On oxidation of hot-pressed $Al_3O_3 - ZrO_2 - TiN$ ceramics [2.132] above 1300° C the specimens failed because of strong internal stresses (Fig. 2.56).

Protective scales exhibiting diffusion-controlled oxidation behaviour occurred on specimens if the reaction layers contained enough silica to place the scale composition either in the two-phase, silica-mullite region or on the mullite-rich side of the mullite-alumina two-phase field of the alumina-silica binary phase diagram. The protective nature of the reaction product in these cases (alumina-SiC, mullite-SiC, mullite-MoSi$_2$) persists even though the reaction involves volume increases of more than 20%. The key to this lies in the viscous nature of the coating which contains either free silica or large amounts

Fig. 2.56. Specimen of the $Al_2O_3 - ZrO_2 - TiN$ composite after heating in air up to 1400° C at a rate of 10° C/min

of mullite. Stresses due to volume expansion are dissipated through viscous flow or creep and, therefore, buckling and coating disruption are avoided.

The oxidation process of SiC-containing ceramics is described, as a rule [2.127], by the parabolic law, but the oxidation rate may be much higher than for pure silicon carbide ceramics [2.130]. The oxidation of SiC whiskers in the Al_2O_3 matrix reveals the formation of graphite [2.129] along with SiO_2. We observed a similar effect on B_4C oxidation (Sect. 2.4), i.e. the oxidation of whiskers proceeds by the reaction

$$SiC + O_2 = SiO_2 + C \quad .$$

SiO_2 interacts with the matrix by the reaction

$$2\,SiO_2 + 3\,Al_2O_3 = 3\,Al_2O_3 \cdot 2\,SiO_2$$

with the formation of mullite.

The interaction of oxidation products of the reinforcing phase with the matrix at high temperatures (1500° C) was also observed for $Al_2O_3 - TiC$ and $Al_2O_3 - MoSi_2$ composites [2.130]. The total overall oxidation reactions for these composites can be written in the following way:

$$2\,TiC + 3\,O_2 + 2\,Al_2O_3 = 2\,Al_2TiO_5 + 2\,CO$$

and

$$2\,Al_2O_3 + MoSi_2 + 7\,O_2 = 3\,Al_2O_3 \cdot 2\,SiO_2 + MoO_3 \quad .$$

The oxidation mechanisms for mullite matrix composites are similar to those of Al_2O_3 composites. The activation energy for the mullite–SiC composite is 376 kJ/mole, which is somewhat lower than for the alumina matrix ceramics [2.130]. The presence of impurities accelerates oxidation [2.130, 133], as it was observed for nonoxide ceramics (Sect. 2.2). The presence of ZrO_2 in the mullite matrix also deteriorates oxidation resistance [2.133].

Thus, the oxidation behaviour of composite systems is influenced not only by the oxidation product of the oxidizable phase, such as silica in the case of either SiC or $MoSi_2$, but also by the reaction of the oxidation product with the matrix material to produce the final coating composition. Further parameters to be considered include volume changes during reaction, thermal expansion mismatch between the coating and the substrate, which is critical to thermal cycling, and rigidity of the coating, which will influence the ability of the coating to accommodate mismatch stresses.

If alumina is used as a matrix for a dispersed, oxidizable phase of silicon such as SiC particles, whiskers or fibers, it is important that silicon be present in sufficient amount to convert all the alumina in the oxidation scale to mullite. Too little silica will result in the formation of a rigid scale which will delaminate and provide no protection. The published results on the oxidation of SiC-whisker-reinforced $Al_2O_3 - ZrO_2$ composites [2.134] are consistent with the data reported for $Al_2O_3 - SiC$ and mullite–SiC composites. The oxidation rates measured in all of these materials are almost identical.

This chapter is concluded with Table 2.11, which summarizes comparative data on the thermal stability of different ceramics and helps to choose the material most suitable for operation in a required temperature range. However, the reader should use these data with some caution since there are many factors than can influence the properties, such as purity, porosity etc.

Table 2.11. Maximum working temperatures of ceramics, ° C

Material	Temperature of fusion or decomposition	Beginning of the oxidation	Maximum working temperature		
			In oxidizing atmospheres		In reducing or inert atmospheres
			Short-term	Long-term	
Silicon nitride	1900				
RBSN		700	1500	1400	1800
HPSN		1000	1500	1400	1600
Silicon carbide	2540				
SiSiC sintered		1000	1650	1400	1400
or HP SiC		1000	1650	1550	2300
Aluminum nitride	2450	800	1500	1400	
Boron carbide	2450	600	1100	600	2000
Boron nitride	2300	700	1100	700	2200
Alumina	2050		1900	1900	1900
Zirconia	2550		2200	2200	2200

3. Gaseous Corrosion of Ceramics

In the previous chapter the behaviour of ceramics on heating in oxygen and air was discussed. However, actual heat engine or heat exchanger environments often contain other oxidants apart from pure oxygen, such as chlorine and sulfur dioxide, as well as condensed phase deposits (Table 3.1).

Table 3.1. Corrosive applications for ceramics [3.1]

Application	Temperature, °C	Pressure, atm	Atmosphere	Deposit
Heat engines	900–1400	1-50	Oxidizing	Na_2SO_4, $Na_2V_xO_y$
Coal combustion	1200–1400	1-10	Reducing	Acidic or basic coal slags
Industrial furnaces	1000–1600	~ 1	Oxidizing, reducing	NaCl, NaF, Na_2SO_4, transition-metal oxides
Magnetohydro-dynamics	1000–1400	1-10	Oxidizing, reducing	K_2CO_3, K_2SO_4
Fuel cells	800–1000	~ 1	Oxidizing	Alkali carbonates

3.1 Hot Corrosion

This section focuses on the oxidation processes on the ceramic surface in the presence of condensed phase deposits. The oxidation of materials under such conditions is usually called hot corrosion. Sodium sulfate and vanadia are the main corrosive deposits encountered in combustion applications. Sodium sulfate decomposes reacting with some reducing agent to form Na_2O, which interacts with V_2O_5 by the reaction

$$m\,Na_2O \cdot n\,V_2O_5 = m\,Na_2O \cdot (n-p)\,V_2O_5 \cdot p\,V_2O_4 + \frac{1}{2}p\,O_2 \quad .$$

Such a melt ingresses at the material/oxide interface and evolves oxygen, since it functions as a kind of oxygen carrier from the atmosphere to the material. If V_2O_5 and Na_2SO_4 themselves are stable, the latter decomposes in the presence of V_2O_5 to form Na_2O and SO_3 [3.2]. The temperature of the Na_2SO_4 dissociation in the presence of, e.g., SiO_2 goes down to 1030° C and that of V_2O_5 down to 740° C. This explains why the $Na_2SO_4 - V_2O_5$ mixture is more corrosive than $Na_2O - V_2O_5$, as it was observed in the study of metals. In the first case, apart from the oxygen evolution, SO_2, which has a much higher oxidability than air, is formed . This is why quite a number of studies deal with the corrosion of ceramics in the stream of fuel products. [3.1] gives a detailed overview of investigations in the field of hot corrosion of silicon carbide and silicon nitride ceramics. Relevant information on hot corrosion of Si_3N_4, SiC and oxide ceramics can also be found in [3.3]. Therefore, in this chapter we only briefly examine the processes occurring on hot corrosion of ceramics.

3.1.1 Silicon Nitride Ceramics

RBSN exhibits high corrosion resistance in the stream of combustion products of pure fuel ($< 10^{-5}\%$ Na and V, 0.5% S) below 1400° C; it corrodes severely when fuel contains 0.005% Na, 0.005% V and 3% S [3.4]. When the specimens are kept for 225 h at 900° and 1000° C in combustion products of such fuel, they are covered with an oxide layer which is ~ 1 mm thick and contains 85% SiO_2, 7% V_2O_5 and 8% Na_2O. Under the same conditions the oxide layer formed on HPSN is only half as thick. The X-ray analysis of the surface layer of specimens revealed Na_2SO_4 after corrosion at 900° C and $Na_2Si_2O_5$ and SiO_2 after corrosion at 1100° C. A protective silica film formed in the initial period on the specimen surface at 1400° C apparently resulted in much weaker corrosion.

Higher corrosion resistance of HPSN in the stream of combustion products compared with reaction-bonded materials was also reported by other researchers. HPSN specimens (Norton Co.) exposed for 7276 h at 870° C to the stream of combustion products of fuel containing 0.0079–0.0116% Na, 0.0036–0.0202% K as well as some sulfur and magnesium did not display any noticeable damage of their surface [3.5]. By thermodynamic calculations a protective SiO_2 film in the presence of sulfur and sodium fails in a reducing atmosphere at relatively low pressures. For example, at 1.4 MPa and 1100° C active corrosion accompanied with silica dissolution is observed only for Na contents exceeding 0.003% and those of $S \sim 0.1\%$ [3.5]. Such pressures are not typical of gas-turbine engines, but they may occur in boilers and gas generators.

The mechanical behaviour and corrosion resistance of Al_2O_3- and Si_3N_4-based materials were studied in the stream of combustion products of a gas mixture [3.6, 7]. The studies on the materials of the system $Si_3N_4 - Al_2O_3$ at 1430° C (on the wall of a chamber) and a flow rate of 500 m/s demonstrated that at the initial stage a mass gain due to Si_3N_4 oxidation was detected and only after a certain period (~ 5 h) a mass loss was observed caused by the erosive damage of the material. Mass gain–mass loss transition time decreases as an Al_2O_3 content increases.

We tested NKKKM-83 ceramic specimens (Sect. 2.1) of a sufficiently large size in a burner rig in the stream of combustion products of TS-1 kerosene burned in air (gas flow pressure of 110–130 kPa, flow rate of 0.20–0.25 kg/s). Temperature variations in the gas stream around the blowing chamber were within $150° - 220°$ C and along its height they were below 100° C. The specimens were first affected by a stationary gas stream for 50 h at a maximum gas temperature of 1250° C and later for 25 h at 1370° C [3.8]. The investigations of the test specimens demonstrated that oxidation considerably changed the morphology of their surface. There are pits of up to 0.5 mm in diameter

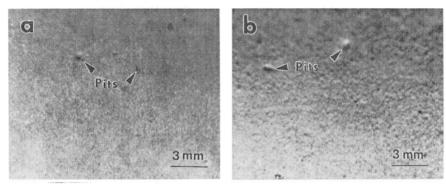

Fig. 3.1. Surface areas of specimens after corrosion in the stream of combustion products (a) and in molten Na_2SO_4 (b)

(Fig. 3.1a) as well as small cracks, pores, white blistered beads. Pores in an oxide layer are stress concentrators and quite often they initiate the cracks. Just the same pits as shown in Fig. 3.1 are formed because of the destruction of grain agglomerates when the specimen surface is etched in molten Na_2SO_4 (Fig. 3.1b).

Thus, the flaws in the surface layer of specimens are, first of all, due to the effect of salt formed on fuel combustion. This can be confirmed by the presence of sodium, sulfur and vanadium traces in the surface layer detected by X-ray microprobe analysis. It also contained several percent of Mg, Ca, Al, Fe, Ba present as impurities in the initial ceramics. The X-ray analysis of the oxide layer revealed α-cristobalite, forsterite, enstatite and an amorphous phase which may include silicates of the above elements.

Light areas (Fig. 3.2b) correspond to compounds of heavier elements, i.e. silicates of the above metals, and dark ones correspond to lighter elements, i.e. silica. A smooth glassy layer on the specimen surface tested in the rig is not formed due to the erosive impact of the gas stream on a liquid oxide layer appearing at test temperatures. A considerable number of flaws in the surface layer of specimens caused some of them to crack during the tests.

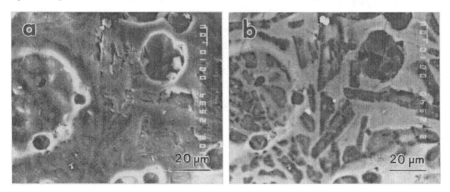

Fig. 3.2. Surface area of a specimen after corrosion in the stream of combustion products in secondary (a) and back-scattered (b) electrons

Table **3.2.** Content of impurities in fuel

Fuel	S	Na	V	Ash content	Mechanical impurities
diesel	0.2	$2 \cdot 10^{-9}$	$2 \cdot 10^{-9}$	0.01	0.005
fuel oil	1.7	$2.3 \cdot 10^{-3}$	$2 \cdot 10^{-3}$	3.25	0.044

The specimens of NKKKM–84 were thermally cycled in the stream of diesel fuel combustion products (Table 3.2) at lower temperatures under two different conditions [3.9]

After the tests under condition 1 the specimens were exposed to a stationary stream of fuel oil combustion products (Table 3.2) at 1050° C for 2 h.

After the tests the specimen surfaces were covered with a black deposit of incomplete fuel combustion products, but no traces of damage were detected. An X-ray analysis of the specimen surfaces revealed only traces of α-cristobalite and forsterite.

To investigate the effect of various substances present in fuel on the corrosion resistance of ceramics, different amounts of sodium, vanadium, and sulfur compounds, as well as sea salts were either added to the fuel [3.4] or deposited on the specimen surface. Studies on the oxidation of specimens with salt solutions deposited on their surface [3.10, 11] can considerably simplify the experiments and give easily interpretable data, since corrosion in combustion products is influenced by too many factors.

RBSN oxidation was studied in the stream of humid and dry air (flow rate 10 ml/min) in the range of 900–1300° C [3.10]. Part of the specimens were oxidized in the initial state, another part after pretreatment with Na_2CO_3 solution and oxidation at 1300° C for 4 h. The initial rate of oxidation is rapid, then it retards abruptly. The activation energy of the process is 130 kJ/mole in the range of 1100–1200° C and 320 kJ/mole below 1100° C. The treatment in the Na_2CO_3 solution noticeably increases the oxidation rate in the initial period. With growing Na_2O content in the surface layer, the mass gain increases due to the formation of low-melting soda-silica glass and acceleration of oxygen diffusion through the oxide layer. Sodium oxide was formed at the expense of the salt decomposition at high temperatures by the reaction $Na_2CO_3 = Na_2O + CO_2$. The presence of sodium sulfate on the surface of high-purity Si_3N_4 specimens decreases the initial temperature of corrosion [3.3]. This may be due to the reaction

$$x\, SiO_2 + Na_2SO_4 = Na_2O \cdot x\, SiO_2 + SO_2 + 1/2O_2 \quad . \tag{3.1}$$

Oxidation in this atmosphere as well as in air leads to the formation of cristobalite. The presence of Na_2SO_4 accelerates the crystallization of amorphous SiO_2 [3.12].

In the case of oxidation at 1400° C the oxide layer also reveals traces of sulfates of impurity elements.

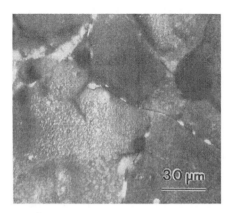

Fig. 3.3. Surface of a specimen oxidized at 1200° C after impregnation with NaCl solution

To simulate the processes occurring on operation of gas turbine blades made of NKKKM–81, the specimens were heavily moistened with solutions of Na_2SO_4, NaCl or sea salt and heated up to 1200° C [3.13]. Figure 3.3 depicts the structure of an oxide layer after salt-assisted oxidation. The X-ray analysis established that it contained α-cristobalite and a glassy phase. The composition and structure of the oxidized layer are virtually independent of the kind of salt deposited on the specimen. All three salts lead to the formation of soda-silica glass on the surface of ceramics. It cracks due to considerable differences in the thermal expansion coefficients of the ceramics and the above layer (Fig. 3.3). Pores in the oxide layer become stress concentrators and initiate the cracks, just as in the case of oxidation in the stream of combustion products.

It is characteristic that the oxidation of ceramics leads to the etching of their surface and the formation of flaws noticeable even despite the oxide layer on the surface of specimens. A similar picture was also reported for NKKKM–84 [3.12]. After dissolution in hydrofluoric acid the oxide layer revealed uncovered numerous pits, whose depth did not exceed their diameter (Fig. 3.4).

Pits are probably formed in the places of SiC inclusions, which are less resistant to sodium salt assisted corrosion [3.1], or due to pore etching in the

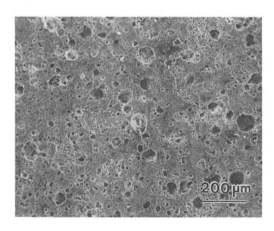

Fig. 3.4. Micrograph of an NKKKM–84 surface after heating up to 1400° C in the presence of NaCl and oxide layer dissolution

ceramics. Studies on the hot corrosion of dense Si_3N_4 produced by different methods [3.14–16] did not reveal a process of pits formation. CVD Si_3N_4 corroded along the grain boundaries, while in the rest of the materials etching of the grain boundary phase was observed. Na_2CO_3 and Na_2SO_4 deposited on the specimen surfaces caused different effects.

The major reason for this is the basicity of Na_2CO_3. The salt easily donates sodium oxide and dissolves the SiO_2 layer. This is a driving force for the rapid decomposition of Na_2CO_3. Thus, in the case of Na_2CO_3 the melt becomes $Na_2O \cdot x(SiO_2)$ at an early stage of the reaction, and Na_2CO_3 does not really play a role. This is in contrast to the case of Na_2SO_4, where Na_2SO_4 is present for a good portion of the reaction. But the reactions by both salts produce important microstructural changes.

Studies on the corrosion of silicon nitride ceramics in Jet A fuel (0.05% S) combustion products discovered separate pits in the surface layer of the specimens [3.16].

We should also note that at the initial stages of oxidation the mass loss proves to be proportional to the amount of salt deposited on the specimen surface [3.15]. This is connected with the decomposition or evaporation of the salt. Both potassium and sodium salts had the same effect [3.17].

The oxidation of RBSN with a porosity of 35% containing $\sim 1\%$ Fe_2O_3 and covered with Na_2SO_4, Na_2CO_3, NaCl, $Na_2V_{12}O_{31}$ and the eutectic mixture Na_2SO_4 − NaCl was investigated at 1000° C for 100 h [3.11].

It is shown that the impregnation of specimens with silicon polymers followed by thermal treatment in nitrogen can reduce their porosity down to 26% and considerably increase corrosion resistance. The protective SiO_2 layer formed under the silicate film on longer oxidation also retards corrosion [3.14].

3.1.2 Silicon Carbide Ceramics

Silicon carbide ceramics are more sensitive to hot corrosion than silicon nitride ones [3.16]. The pitting process was very active in the surface layer of sintered α-SiC specimens on corrosion in the stream of combustion products simulating the operating environment of a gas turbine. The addition of 2 ppm Na to the fuel abruptly increased the corrosion rate. The behaviour of sintered β-SiC and siliconized SiC is not essentially different from that of α-SiC [3.16].

It is shown that the corrosion process of sintered α-SiC in combustion products can be simulated accurately enough by Na_2SO_4-assisted oxidation of specimens. The composition of corrosion products and the condition of the specimens after the tests at 1000° C in the furnace and in the gas stream at 4 atm and a flow rate of 100 m/s were only slightly different. In both cases sodium silicate and pits were formed.

Corrosion of sintered α-SiC (Carborundum Co.) coated with Na_2CO_3 and Na_2SO_4 in air and in 0.1% SO_2/O_2 (for Na_2SO_4) and 0.1% CO/O_2 (for Na_2CO_3) mixtures was studied at 1000° C for 48 h [3.19]. The authors of [3.19] are of the opinion that sodium sulfate hardly interacts with pure SiC. The oxidation of Na_2SO_4-coated specimens was attributed to sodium sulfate decomposition by the reaction $Na_2SO_4 = Na_2O + SO_3$ and to its interaction with SiO_2 formed on SiC oxidation by reaction (3.1) and by the reactions:

$$Na_2SO_4 + SiO_2 = Na_2SiO_3 + SO_3 \quad \text{or}$$
$$Na_2SO_4 + 2\ SiO_2 = Na_2Si_2O_5 + SO_3 \quad . \tag{3.2}$$

However, these reactions are quite improbable from the thermodynamic point of view ($\Delta G > 0$). At the same time sodium sulfate reacts with the carbon present in the material by the reactions:

$$Na_2SO_4 + 2\ C = Na_2S + 2\ CO_2 \quad (\Delta G = -202.2\,kJ/mole)$$

and

$$Na_2SO_4 + 4\ C = Na_2S + 4\ CO \quad (\Delta G = -308.5\,kJ/mole) \quad .$$

The investigation of the $Na_2SO_4 - C$ system demonstrated experimentally the formation of Na_2S. Taking this into account, we can write down the following reactions:

$$SiO_2 + Na_2SO_4 + 2\ C + 3/2\ O_2 = Na_2SiO_3 + SO_2 + 2\ CO_2$$
$$(\Delta G = -665.3\,kJ/mole)$$

and

$$SiO_2 + Na_2SO_4 + 4\ C + 3/2\ O_2 = Na_2SiO_2 + SO_2 + 4\ CO$$
$$(\Delta G = -771.6\,kJ/mole) \quad .$$

The thermodynamic probability of these reactions is very high. Thus, Na_2SO_4-assisted corrosion etches carbon inclusions.

In the case of Na_2CO_3-assisted corrosion the salt reacts actively with the silica formed on oxidation $Na_2CO_3 + x\ SiO_2 = Na_2O \cdot x\ SiO_2 + CO_2$. The formation of sodium silicate, as mentioned above, facilitates oxygen diffusion through the oxide layer and accelerates SiC oxidation. In this case the pitting process in the surface layer was also observed [3.20]. A supposed pitting mechanism is presented in Fig. 3.5.

The effect of the environment on the corrosion of sintered α-SiC is illustrated by Table 3.3 according to the data presented in [3.19].

Electrochemical measurements of Na_2O activity [3.22] have demonstrated that carbon creates basic conditions in Na_2SO_4, which results in a more rapid dissolution of a protective SiO_2 film and accelerated corrosion of ceramics.

Table 3.3. Hot salt corrosion of SiC at 1000° C

Batch	Total amount of corrosion product, mg/cm^2
Na_2SO_4/SO_3	11.65 ± 0.98
Na_2CO_3/CO_2	6.18 ± 0.27
Na_2SO_4/air	5.75 ± 1.30
Oxidation	0.47 ± 0.34

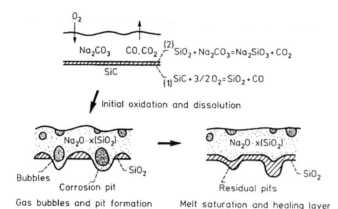

Fig. 3.5. Scheme of proposed pitting mechanism in SiC via bubbles [3.1]

This is a possible explanation for the more active SiC corrosion compared with Si_3N_4.

3.1.3 Oxide Ceramics

Commercial partially stabilized zirconia (Mg-PSZ) exhibits high corrosion resistance in the stream of diesel fuel combustion products at temperatures up to 850° C [3.21]. The surface of specimens was not corrosion-damaged, despite the deposits of combustion products. At the same time the growth of monoclinic ZrO_2 contents in the specimens adversely affects mechanical properties.

Alumina ceramics hardly interact with the $SO_3 - SO_2 - O_2$ mixture at 1000° C and are quite resistant in atmospheres simulating the operating environment of gas-turbine engines [3.3].

3.2 Water Vapour Corrosion

3.2.1 Nonoxide Ceramics

The oxidation rate of silicon nitride in humid air is twice as high as in dry air [3.23]. The interaction of Si_3N_4 with water vapours leads to the formation of ammonia in accordance with the reaction

$$Si_3N_4 + 6\,H_2O = 3\,SiO_2 + 4\,NH_3 \quad .$$

However, even by mass spectroscopy the authors of [3.24] did not manage to detect ammonia in the products of the reaction of Si_3N_4 with humid air (partial pressure of water vapours was 3.3 kPa).

Since ceramics are a promising material for engines using hydrogen fuel, corrosion in the $H_2 - H_2O$ atmosphere has been discussed in a number of recent publications. At relatively low partial pressures P_{H_2O} in H_2 the reaction proceeds [3.25]:

$$Si_3N_4(s) + 3\,H_2O(g) = 3\,SiO(g) + 2\,N_2(g) + 3\,H_2(g) \quad . \tag{3.3}$$

Higher mass losses observed with increasing P_{H_2O} in H_2 support the argument that the corrosion of Si_3N_4 in $H_2 - H_2O$ environments occurs by the active oxidation reaction (3.3). A strong dependency of mass loss on the flow rate of an ambient gas containing a fixed P_{H_2O} was observed. This implies that the transport of the gaseous species through a stagnant boundary layer plays an important role in the corrosion process.

The changes in specific mass and surface morphology of Si_3N_4 made by chemical vapour deposition (CVD) and hot isostatic pressing (HIP) and the changes in strength of $HIPSi_3N_4$ were determined exposing them for 10 h at 1400° C to the $H_2 - H_2O$ atmosphere [3.25]. The corrosion behaviour of both materials depended on the level of oxidant (H_2O) in the atmosphere. Similar trends in mass change were observed in both materials, but the magnitudes of the mass changes in $HIPSi_3N_4$ were significantly greater. This behaviour can be attributed to a larger specific grain-boundary area in the finer grained HIP material, i.e. a larger grain-boundary area was exposed to the corroding gases.

When the water vapour pressure in the H_2 atmospheres was low ($P_{H_2O} < 20$ Pa) mass losses were observed in both materials. As the P_{H_2O} in H_2 was increased, the formation of solid SiO_2 on the surface became thermodynamically favourable, and the mass losses became smaller.

Silicon carbide is oxidized in humid air by the reaction [3.24]:

$$SiC + 2\,H_2O = SiO_2 + CH_4 \quad ,$$

the oxidation rate growing with an increase in the water vapour content in air. Water vapours also accelerate the crystallization of amorphous SiO_2 in an oxide layer [3.26].

Corrosion of SiC in the $H_2O - H_2$ atmosphere led to a mass loss [3.27], just as for Si_3N_4, according to the reactions:

$$SiC(s) + H_2(g) + H_2O(g) = SiO(g) + CH_4(g) \quad , \tag{3.4}$$

$$SiC(s) + 2\,H_2O(g) = SiO(g) + CO(g) + 2\,H_2(g) \quad . \tag{3.5}$$

The mass loss of SiC as a function of time was linear up to several hours.

The thermodynamic probability of different reactions possible on SiC oxidation in the H_2O atmosphere was calculated in [3.28]. It is shown that oxidation in water vapours proceeds by the most probable reaction $SiC + 3\,H_2O = SiO_2 + CO + 3\,H_2$. At low partial H_2O pressures the probability of reactions (3.4, 5) increases.

The corrosion of SiC in hydrogen at temperatures between 1400° C and 1527° C at water vapour pressures between 10^{-6} and 10^{-3} MPa occurs by active oxidation of SiC by the water vapour with the formation of SiO and CO. At low water vapour pressures, the transport of H_2O from the gas stream to the surface controls the reaction rate. At high water vapour pressures, SiO_2 is present as discrete particles on the surface and continued active oxidation occurs with the concurrent reduction of this SiO_2 [3.27]. It is also established that first, the rate of corrosion increases almost linearly with water vapour pressure, reaches a maximum, and then decreases, second, the maximum in the corrosion rate shifts to lower water vapour pressures as the temperature decreases, third, at low water vapour pressures, the corrosion rate is weakly dependent on temperature. Conversely, the corrosion rate is strongly dependent on temperature at high water vapour pressures.

The oxidation of sintered α-SiC with boron and carbon in the H_2–H_2O–Ar atmosphere at 1300° C under active oxidation conditions revealed extensive grain-boundary attack [3.29].

The corrosion of aluminium nitride (3% porosity) with 1% CaO sintered at 1800° C was studied in the hydrogen stream containing 5%–70% of water vapours [3.30]. As the experiments have shown, the kinetic curves of corrosion in humid hydrogen at 1300° and 1450° C obey the parabolic law in the initial period and then become linear. The corrosion products are α-Al_2O_3 and $CaO \cdot 6Al_2O_3$. The latter was formed as a result of the interaction between the secondary grain-boundary phase with the composition $CaO \cdot 2Al_2O_3$ and Al_2O_3. The apparent activation energy calculated by the linear section of the curves is $231 \pm 25 \, kJ/mole$, while on oxidation in oxygen it is equal to $269 \pm 30 \, kJ/mole$. This difference in activation energies can be attributed to the fact that the Al–N bond breaks more easily when interacting with H_2O than with O_2 [3.30]. Though in [3.11] water vapour is reported to accelerate the oxidation of HPAlN, the process obeys the parabolic law. Up to 1250° C the reaction between AlN and adsorbed H_2O, and above 1300° C the diffusion of moisture through Al_2O_3 are the limiting stages of the process.

The interaction of boron carbide with water vapours starts already at 250° C. Here the following reactions are possible:

$$B_4C + 8 H_2O = 2 B_2O_3 + CO_2 + 8 H_2 \quad ,$$

$$B_4C + 6 H_2O = 2 B_2O_3 + C + 6 H_2 \quad ,$$

$$B_2O_3 + H_2O = 2 HBO_2, \quad B_2O_3 + 3 H_2O = 2 H_3BO_3 \quad .$$

Gaseous boric acid removes a boron oxide film. At 550°–600° C with a dew point of 25°–70° C and at 650° C with a dew point of 88° C the rates of formation and removal of the B_2O_3 film are equal. At higher temperatures B_2O_3 is formed at a higher rate than it is removed because of the interaction with water vapours. Therefore, at low temperatures boron carbide is oxidized with water vapours more rapidly than with dry air, at high temperatures the situation is quite the opposite [3.9].

Boron nitride is hydrolysed with humid air to form ammonia and boric acid: $BN + 3 H_2O = H_3BO_3 + NH_3$.

3.2.2 Zirconia Ceramics

Corrosion of ZrO_2 ceramics in wet environments is quite a special case.

This material is subjected to so-called low-temperature degradation at $200°-300°$ C in humid air or at lower temperatures in water. The following phenomena characterizing this degradation have been observed [3.32–36]:

1) The degradation occurs significantly in a specific temperature region, at $200°-300°$ C in air.
2) The degradation proceeds from the surface to the interior of the specimen.
3) Micro- and macrocracking occur because of the spontaneous phase transformation from tetragonal to monoclinic.
4) The presence of water accelerates the transformation and the degradation.
5) Higher water vapour pressures accelerate the transformation.
6) The degradation can be prevented by increasing dopants or by reducing grain sizes.
7) H_2O treatments bring about a mass increase and a lattice expansion.
8) Mass and lattice revers to their original values by heating in air or in vacuum.
9) The infrared spectra demonstrate the formation and the diminishing of OH^- by H_2O treatment and subsequent reheating.

Several models have been proposed to explain these phenomena. *Sato* [3.32] reported that the chemisorption of H_2O to form OH^- ions at the surface resulted in the loss of constraint energy, which led to the transformation. *Lange* [3.33] supposed that the destabilization of tetragonal zirconia polycrystals (Y-TZP) by H_2O occurred due to the formation of $Y(OH)_3$ crystals at the surface.

The authors of [3.35] believe that structural degradation of TZP ceramics in water and water vapour at low temperatures is caused by grain boundary dissolution, which relieves the internal stresses and thereby can induce the phase transformation to monoclinic structure. The dissolution rate depends on the defect concentration of the structure as well as on the stabilizers used and their influence on vacancy formation. Y_2O_3 and Al_2O_3 doping increases the dissolution rate by the vacancy formation mechanism, but decreases the phase transformation rate by stabilizing the tetragonal structure. CeO_2 doping improves the aging properties of TZP ceramics by stabilizing the tetragonal structure and not affecting the dissolution rate of the grain boundaries. *Yoshimura* [3.36, 37] supposed that corrosion involved the introduction and exclusion of OH^- ions into and from the Y-TZP lattice. The following steps of a degradation mechanism of Y-TZP by H_2O, as shown in Fig. 3.6, have been proposed:

Stage 1: chemisorption of H_2O at the surface
Stage 2: formation of Zr–OH or Y–OH bonds which bring about the lattice strain at the surface
Stage 3: the strain accumulates by the diffusion of OH^- at the surface and in the lattice

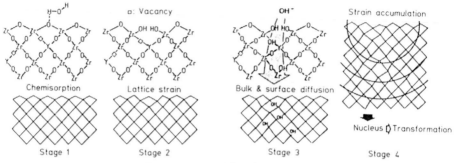

Fig. 3.6. Degradation process of Y–TZP by H_2O [3.37]

Stage 4: the accumulated strain area can act as a nucleus of the monoclinic
phase in the tetragonal matrix.

The phase transformation from tetragonal to monoclinic causes an increase
in cell volume, which results in the cracking.

One of the most important features in this model is the accumulation of
strains, because it biases the activation energy which corresponds to the energy
barrier from tetragonal to monoclinic. When the accumulated strain exceeds
a critical value, the activation energy becomes *approx*0, i.e. barrierless, which
induces the spontaneous nucleation of a monoclinic domain in the tetragonal
matrix.

3.3 Corrosion in Carbon Oxide Environments

Carbon oxides are formed on the combustion of various fuels. Therefore numerous materials are affected by these gases at high temperatures. Corrosion
in carbon oxide environments has already been discussed in Sect. 3.1, but only
for the case of Na_2CO_3 present on the specimen surface. However, carbon-
containing gas atmospheres themselves can interact with ceramics.

As is shown in [3.38], at high temperatures silicon nitride ceramics exhibit
a higher resistance in the CO/CO_2 mixtures at a partial oxygen pressure of
0.1–10 Pa than in oxygen or in air. At an oxygen pressure of 1.33 Pa such a
mixture caused a mass loss due to the formation of gaseous silicon monoxide.

When Si_3N_4 is heated in the CO atmosphere, the following reactions
proceed: $Si_3N_4 + CO = Si_2N_2O + SiC + N_2$ between 1450° and 1550° C;
$2\ Si_2N_2O + CO = 3\ SiO + SiC + 2\ N_2$ above 1500° C; $2Si_3N_4 + 3\ CO =
3\ SiC + 3\ SiO + 4\ N_2$ at 1750° C. The addition of CaF_2 and MgO promotes the
conversion of Si_3N_4 into silicon oxynitride in the CO atmosphere at 1500° C. At
1400°–1500° C silicon nitride in a coke charge decomposes to form oxynitride
and cubic silicon carbide [3.39].

The oxidation of SiC with CO_2 starts at 1000° C, just as the oxidation in
air, to form SiO_2: $SiC + 3\ CO_2 = 4\ CO + SiO_2$. As can be observed, the kinetic

curves plotted for SiC single crystals and polycrystalline self-bonded SiC at 1500° and 1650° C are in good agreement [3.39].

The corrosion of SiC designed for heat exchanger components was investigated on a lab scale simulating coal gasification conditions [3.40]. Silicon carbide ceramics are shown to be resistant in carbon oxide environments at high temperatures, but they corrode when affected by melted slag formed during coal gasification.

High-temperature (1700°–2000° C) corrosion of sintered SiC in the flame of an acetylene burner (mixture of H_2O, CO, H_2, O_2 and C_xH_y) results in the burn-out of carbon inclusions and in the mass loss of specimens due to SiO formation [3.41].

Comparing the data cited in Sects. 2.2, 3.2 and in this section, one can conclude that under low-temperature oxidation conditions the chemical activity of gases for SiC increases in the following order: $H_2O \rightarrow CO_2 \rightarrow O_2$. At high temperatures the situation will be quite the opposite.

The surface of AlN heated in the carbon monoxide atmosphere is covered with an Al_2OC film which protects nitride from a further CO effect.

At 700°–800° C boron nitride is oxidized in CO_2, just as in air, with the formation of B_2O_3 and nitrogen.

When boron nitride interacts with gas streams of high flow rates at 6000°–7000° C and atmospheric pressure, it erodes rapidly in air and not so rapidly in exhaust gases and nitrogen. The analysis of the obtained data shows that higher erosion in the streams of air and fuel combustion products is determined by oxidation only to a small extent; it is mainly caused by mass and heat transfer processes accounting for 75% of erosive nitride decomposition. The kinetics of boron nitiride dissociation by the reaction $2 \, BN = 2 \, B + N_2$ and its release can unambiguously be attributed to a heat load and a heating rate in inert gas atmospheres and in streams [2.39].

3.4 Corrosion in Halogen- and Chalcogen-Containing Environments

It is reported in [3.39] that chlorine weakly affects silicon nitride at average temperatures. Thus, at 350°–420° C the mass loss is 0.77%–0.95% after 2 h treatment of powder with chlorine, while the calcination of silicon nitride powder in the stream of chlorine for 1 h at 1200° C [3.23] results not in a loss but in a mass gain by 1%. HPSN and CVD Si_3N_4 are also highly resistant to the effect of 1%–2% Cl_2/10%–20% O_2/Ar mixtures [3.42, 43], though thermodynamic calculations predict the formation of $SiCl_4$ on the interaction of Si_3N_4 with a mixture of chlorine and argon [3.42]. The absence of interaction is explained by the formation of a protective SiO_2 or Si_2N_2O layer. Pure SiO_2 does not react with chlorine at 950° C [3.42]. Halogens react with SiC to form mainly silicon compounds. Thus, the reaction with chlorine $SiC + 2 \, Cl_2 = SiCl_4 + C$ proceeds already at 100° C; at 500° C they start to react actively.

With an increase in temperature up to 1000° C the following reaction proceeds:

$$SiC + 4\,Cl_2 = SiCl_4 + CCl_4 \quad .$$

The etching rate with chlorine at 1000° C is 0.5 μm/min. A mixture of chlorine and oxygen increases this rate up to 1 μm/min.

As is reported in [3.44], small amounts of oxygen in chlorine accelerate corrosion, large amounts cause the active–passive oxidation transition due to a protective SiO_2 layer.

Mass spectrometry [3.43,45] reveals $SiCl_4$, $SiCl_3$ and oxychlorides Si_2OCl_6 and Si_3OCl_8 formed in the oxygen/chlorine mixture according to the reactions:

$$4SiCl_4 + O_2 = 2Si_2OCl_6 + 2Cl_2 \quad ,$$

$$Si_2OCl_6 + SiCl_4 = Si_3OCl_8 + Cl_2 \quad .$$

Self-bonded SiC (NC–430, Norton Co.) is less corrosion-resistant in chlorine-containing environments due to the etching of silicon inclusions. However, the addition of oxygen to the gas mixture also results in the passivation of this material. Hot-pressed NC–203 and SiC single crystals exhibit a rather high corrosion resistance [3.42]. A limiting stage of the reaction with the $Cl_2/O_2/Ar$ mixture, as proposed in [3.43], is the diffusion of silicon chlorides through the SiO_2 layer. In [3.46] it is suggested that the solubility of silicon chlorides in SiO_2 is lower than Cl_2, therefore at 1300° C a liquid oxide layer becomes blistered. At 1300° C the reaction rate reaches a maximum at a high partial oxygen pressure, as opposed to 900°–1100° C [3.44]. The addition of hydrogen to chlorine lowers the corrosion rate.

At relatively low temperatures SiC also reacts with ClF_3. The interaction of SiC with fluorine is the most active to form SiF_4 and CF_4.

The corrosion of SiSiC in air with the addition of 2 vol.% H_2O and 0.1% HF is most active at 500°–700° C [3.47].

By thermodynamic calculations SiC has to interact with sulfur to form SiS_2 and carbon, while it does not necessarily interact with selenium and tellurium [3.39].

Aluminium nitride starts to react with chlorine above 700° C forming $AlCl_3$, but dry hydrogen chloride does not affect it. In sulfur, carbon disulfide and phosphorus vapours AlN decomposes partially, in sulfur dichloride vapours it decomposes totally. PCl_3 does not affect aluminium nitride [3.39].

Boron carbide does not interact with sulfur and phosphorus vapours, as well as with nitrogen up to 1200° C; with chlorine it reacts above 1000° C to form BCl_3 and graphite. Bromine and iodine do not react with B_4C [3.39].

Boron nitride is resistant to hydrogen and sulfur dioxide; it reacts with fluorine at room temperature by the reaction $2BN + 3F_2 = 2BF_3 + N_2$. The absence of interaction with other halogens and some halides is a basis for the removal of boron, boron carbide and other impurities from boron nitride [3.39].

Table 3.4. Resistance of ceramics to attack by gases and vapours [3.48]

Medium	Silicon nitride	Silicon carbide	Boron nitride	Boron carbide	Alumina	Zirconia, stabilized
Hydrogen		A<1430 C>1430	A	B1200	A1700	C
Ammonia	A	A	A3000		A	C>2200
Nitrogen	A1800	A1100 B1400	A	A1200	A1700	B2200
Argon	A	A2000	A	A2000	A1700	A
Water vapour·	A800	B1150	C220	A260	A1700	B200 C1800
Carbon monoxide		B1300	A800	A2200	A1700	A1400
Carbon dioxide		A1000		A1100	A1200	A1200
Sulphur dioxide		A1050			A	A
Hydrogen sulphide	A1000			B	B	
Hydrogen chloride		A1200	B		A	A
Fluorine	C	C1400	C	C	A	C
Chlorine	A900	A<700 C>700	A700	B700	A	A

Key:
A - resistant to attack up to temperature indicated (°C),
B - Some reaction at temperature indicated,
C - Appreciable attack at temperature indicated

Hot-pressed boron nitride interacts with chlorine very slowly at 700° C, but rapidly enough at 1000° C to form BCl_3 [3.39].

The heat treatment of Y–TZP in CF_4 or CF_2Cl_2 gases at 300°–500° C yielded some doping of fluorine in the surface. Here the formation of F-doped tetragonal and cubic phases is possible [3.37].

This chapter is concluded with Table 3.4 which summarizes data on the corrosion resistance of different ceramics and helps to choose the material most suitable for operation in a required gas atmosphere. The results in this table should be assessed in conjunction with the remarks given in the relevant sections of Chap. 3.

4. Corrosion in Liquid Media

Ceramics used for different applications are affected not only by gas atmospheres, they can also be exposed to liquids, viz. solutions and melts. Corrosion of metals in liquid media is usually subdivided into chemical (in nonelectrolytes and molten metals) and electrochemical (in electrolytic solutions and melts). Since the corrosion mechanisms of ceramic materials in liquid media have not yet been studied thoroughly enough, we use basic chemical laws to describe the process. Under real conditions the processes of corrosive damage of structural ceramics, especially conducting SiC- and B_4C-based ones, can also be described from the point of view of electrochemistry using, e.g., the theory of microgalvanic cells by Akimov.

In this chapter we examine the corrosion of ceramics in alkaline solutions and acids as well as in molten salts and alkalies. We will not dwell on the corrosion in molten metals, since this problem is treated in detail in the literature devoted to refractory ceramics used in metallurgy.

4.1 Corrosion in Solutions

Corrosion resistance of ceramics in different solutions determines their applicability for manufacturing components of chemical equipment, acid pumps, bearings operating in aggressive environments, etc.

4.1.1 Silicon Nitride Ceramics

Powders and, recently, dense materials have been used to study the corrosion resistance of silicon nitride in liquid media. Just as corrosion resistance in other environments, it depends on purity and particle sizes of powders and on the method of producing compact specimens, i.e. on their density and purity.

According to [4.1], silicon nitride powders are resistant to attack by sulfuric, hydrochloric, nitric, $m-$, $o-$ and pyrophosphoric acid solutions of any concentration as well as by aqua regia. Phosphoric and hydrofluoric acids decompose them only partially on heating. The interaction with phosphoric acid starts above 100° C and up to 200° C proceeds rather slowly. At higher temperatures the reaction rate grows abruptly. The reaction product is $SiO(PO_4)_4$ converted into low-soluble phosphate $Si_3(PO_4)_4$ at 300° C. Concentrated sulfuric acid decomposes Si_3N_4 on longer heating to form ammonium sulfate. Thus, upon

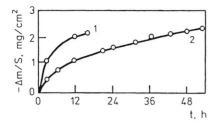

Fig. 4.1. Kinetics of dissolution of HPSN with 10% MgO and 1% CaF_2 (1) and with 10% MgO (2) in a boiling 14% H_2SO_4 solution [4.2]

heating for 3–6 h, 42.1%–45.8% nitrogen evolves. At the same time concentrated boiling solutions of $CuSO_4$ and $KHSO_4$ in H_2SO_4 do not affect silicon nitride for over 500 h. Silicon nitride dissolves in hydrofluoric acid more actively; however, in the mixtures with other acids, except HNO_3, the rate of the process decreases. Silicon nitride also interacts with fluorides such as NH_4F, HBF_4 and others.

Silicon nitride exhibits a higher resistance in alkaline solutions than in melts (Sect. 4.2). Its resistance increases with decreasing concentration, and in 50% solutions it does virtually not decompose. Silicon nitride whiskers possess a higher resistance than powders.

Silicon nitride materials used for manufacturing friction surfaces which operate in aggressive environments have to contain antifriction agents such as CaF_2. Hot-pressed materials with 10% MgO and 1% CaF_2 corrode slowly in a 14% H_2SO_4 solution (Fig. 4.1). Higher porosity of materials results in lower corrosion resistance. The corrosion of materials pressed without sintering aids and having a porosity of 20% is slightly more active [4.3]. The mass loss of RBSN with 5% ZrO_2 and of NKKKM-80 for 1000 h in H_2SO_4 at room temperature did not exceed 1% [4.4].

The detectable mass loss of specimens in acids is apparently not caused by the decomposition of Si_3N_4 but by that of additives and impurities present in the materials. When the amount of CaF_2 added to HPSN increases from 1% to 20%, the corrosion rate becomes almost two orders higher [4.2]. This can be explained by the interaction of CaF_2 with sulfuric acid, by the reaction $CaF_2 + H_2SO_4 = CaSO_4 + 2HF$ and by further dissolution of Si_3N_4 in hydrofluoric acid:

$$Si_3N_4 + 16HF = SiF_4 + 2(NH_4)_2SiF_6 \quad .$$

Materials containing less than 1% CaF_2 exhibit satisfactory corrosion resistance in strong acids (H_2SO_4, HNO_3 etc.) [4.2, 3].

The leaching behaviour of HPSN containing Y_2O_3, Al_2O_3 and AlN as additives and HIPSN without additives were studied in 0.1 to 10 M HF aqueous solutions at 50°–80° C [4.5]. While silicon and aluminium ions were dissolved into HF solutions, yttrium ions did not dissolve at all and formed insoluble YF_3. The dissolution of silicon and aluminium ions was controlled by the surface chemical reaction, and the apparent activation energies were 70.5 to 87.6 kJ/mole, respectively. The corrosion rate increased with an increasing degree of crystal-

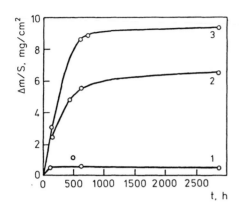

Fig. 4.2. Mass gain of NKKKM–83 ceramics as a function of time of exposure to o–phosphoric acid with a concentration of 10 (1) 50 (2) and 100% (3)

lization of the grain-boundary phases. The corrosion resulted in roughness of the surface and strength degradation. Si_3N_4 ceramics containing an amorphous phase at the grain boundaries showed the most excellent resistance to corrosion with the HF solution, and kept a fracture strength of above 400 MPa even after leaching 40% of the silicon ions.

The studies on the resistance of NKKKM–83 ceramics (Sect. 2.1.4) to attack by o–phosphoric acid solutions at room temperature have demonstrated [4.6] that the mass gain of the specimens grows with its concentration (Fig. 4.2). The interaction of phosphoric acid with magnesium silicates present in ceramics can result in a number of products, including insoluble silicon phosphate and magnesium orthophosphate. The layer of insoluble corrosion products formed on the surface and in the pores of the specimens abruptly decreases the rate of the process after 500 h exposure. Subsequently, hydrophosphates with a good solubility in aqueous solutions can be formed on the specimen surface. The X-ray analysis did not succeed in establishing the composition of the deposits revealed on the specimen surface after corrosion.

HPSN [4.3] and RBSN are also stable enough in alkaline solutions. The mass loss in a 20% NaOH solution at 20° C for 1000 h was 0.12% for RBSN with 5% ZrO_2, 0.24% for NKKKM–80 and 0.65% for NKKKM–83 [4.4].

The electrochemical oxidation of CVD Si_3N_4 films on silicon was studied in [4.7]. The anodic oxidation proceeds by the reaction $6O^{2-} + Si_3N_4 = 3SiO_2 + 2N_2 + 12e^-$. The conversion rate remains constant in the range of current density of 0.4–1.3 A/m^2. The content of residual nitrogen in the intermediate product (silicon oxynitride) decreases with an increase in current density. Further oxidation leads to the conversion of oxynitride into silica. Silicon oxynitride is less resistant to acid and alkali attack than nitride [4.1].

4.1.2 Silicon Carbide Ceramics

Silicon carbide powder, just as that of nitride, does not dissolve in hydrochloric, sulfuric and nitric acids. As opposed to Si_3N_4, it does not react with hydrofluoric acid [4.8]. The data on SiC resistance in an HNO_3 and HF mixture are quite contradictory. Some authors are of opinion that the chemical resistance in this mixture is determined by the presence of different polytypes. According to [4.9], SiC powder with particle sizes of up to 15 μm is resistant to attack by the HNO_3 and HF mixture irrespective of polymorphic modification and polytypic composition. However, at 200°–250° C phosphoric acid decomposes SiC with the release of SiO_2, CO_2, H_2 and CH_4. Dense silicon carbide materials also display a good resistance to attack by different mineral acids and their mixtures. They virtually do not dissolve on boiling in a 10% Na_2SO_4 solution; they exhibit rather high activity interacting with a 10% Na_2CO_3 solution and even higher activity interacting with concentrated alkaline solutions [4.8].

The mass loss for recrystallized silicon carbide of AnnaNox CK and Crystar grades exposed to a 20% NaOH solution for 1000 h at room temperature makes up 0.26%–0.27% and 0.45%–0.49% on attack by concentrated sulfuric acid [4.4].

Self-bonded silicon carbide containing 5-15% of free silicon is less stable in alkaline solutions. This is due to the reaction between free silicon and NaOH: $Si + 2NaOH + H_2O = Na_2SiO_3 + 2H_2$. A white deposit sometimes covering the specimen surface after the tests results from silicate hydrolysis proceeding by the reaction $SiO_3^{2-} + 2H_2O = H_2SiO_3 + 2OH^-$. The XRD analysis could not reveal free silicon in self-bonded SiC specimens exposed to a 20% NaOH solution for 1000 h [4.4]. They acquire a complex system of bonded pores. The total mass loss is $\sim$ 13%. The data on the resistance of self-bonded silicon carbide materials of Japanese production to attack by KOH and NaOH solutions are presented in Fig. 4.3. In a 40% K_2CO_3 solution at 40° C these materials corrode to a less extent [4.10].

Free silicon in the composition of self-bonded materials also interacts easily with hydrofluoric acid by the reaction $Si + 6HF = H_2SiF_6 + 2H_2$. At the same time self-bonded SiC is highly resistant to attack by sulfuric and hydrochloric acids and can successfully be used, e.g., for manufacturing the components of acid pumps [4.8].

At room temperature the $SiC - TiB_2$ composites are resistant to attack by 50% NaOH, a 10% HF+57% HNO_3+33 % H_2O mixture and aqua regia. After

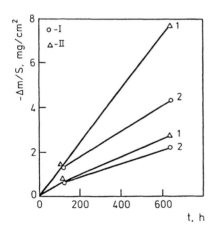

Fig. 4.3. Dissolution kinetics of self-bonded SiC produced by Nippon Tungsten (I) and S Co. (II) (Japan) in a 43% KOH solution (1) and a 48% NaOH solution (2) at 20° C [4.10]

the exposure to these solutions the properties of the specimens changed only inconsiderably [4.11].

The applicability of different SiC materials for manufacturing face seals was estimated in [4.12]. It was shown that sintered α-SiC exhibit the highest resistance compared to self-bonded SiC, sintered β-SiC, siliconized graphite and other materials in the following solutions: 30% oleum, 98% H_2SO_4, 53% HF, 70% HNO_3, 10% HF+57% HNO_3, 25% HCl, 85% H_3PO_4, 45% KOH and 50% NaOH up to 100° C and can operate in these media. Self-bonded SiC corrodes noticeably, e.g., in alkalies and 30% oleum. Mass loss in phosphoric and hydrofluoric acid solutions is small, but the corrosion badly deteriorates its mechanical properties. The resistance of silicon carbide materials exposed to different acid solutions (e.g. H_2SO_4) can be improved by their preoxidation leading to the formation of a continuous silica film on the surface [4.13].

In [4.14] corrosion in water saturated with oxygen and oxygen-free was investigated at 290° C. The presence of oxygen proved to accelerate the corrosion of silicon carbide ceramics. The process concentrated along the grain boundaries and around pores and was accompanied with silica sol formation.

The comparative data on the corrosion resistance of sintered α-SiC, self-bonded SiC and Al_2O_3-based materials are summarized in Table 4.1.

Table 4.1. Corrosion resistance of different materials* [4.12]

Corrosive medium	T, ° C	$-\Delta m$, mg/(cm^2 · y)		
		sintered SiC	self-bonded SiC (12% Si)	dense alumina (99% Al_2O_3)
Solution				
98% H_2SO_4	100	1.8	55.0	65.0
50% NaOH	100	2.5	< 1000	75.0
53% HF	25	< 0.2	7.9	20.0
85% H_3PO_4	100	< 0.2	8.8	< 1000
70% HNO_3	100	< 0.2	0.5	7.0
45% KOH	100	< 0.2	< 1000	6.0
25% HCl	70	< 0.2	0.9	72.0
10% HF 57% HNO_3	25	< 0.2	< 1000	16.0

* Test time 100–300 h

4.1.3 Aluminium Nitride Ceramics

Aluminium nitride powder, as opposed to other examined nonmetallic refractories, hydrolyses in boiling water quite easily according to the reaction AlN + 3H$_2$O = Al(OH)$_3$ + NH$_3$. In [4.15] γ-AlOOH is reported to form when AlN is boiled in water.

However, high-purity AlN is much more resistant than commercial AlN. Aluminium nitride exhibits higher resistance in concentrated sulfuric, hydrochloric and nitric acids than in diluted ones. At 120° C 10%–40% NaOH

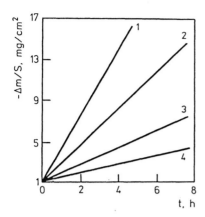

Fig. 4.4. Dissolution kinetics of sintered AlN at 20° C in 2 M HCl (1), 1 M H_2SO_4 (2), 2 M HNO_3 (3), and 2 M CH_3COOH (4)

solutions dissolve aluminium nitride almost entirely within 1 h. As is shown in [4.16], on the interaction with HCl the most negative change of Gibbs energy is characteristic of the reaction $AlN + 4HCl = AlCl_3 + NH_4Cl$. The activation energy of sintered AlN dissolution in 2 M HCl is 50 kJ/mole in the range of 0°–60° C.

Aluminium nitride powder heated in diluted sulfuric acid decomposes completely within 4 days according to the reaction $2AlN + H_2SO_4 + 6H_2O = 2Al(OH)_3 + (NH_4)_2SO_4$. Sintered AlN specimens are more resistant to acid attack than powders (Fig. 4.4). However, AlN specimens produced by carbothermal method and $AlCl_3 \cdot NH_3$ decomposition interact with alkalies just as easily as powders (Table 4.2).

Table 4.2. Decomposition rate of AlN specimens in a KOH solution [4.16]

Preparation method	C, mole/l	T, ° C	$-\Delta m$, mg/(cm$^2 \cdot$ h)
$AlCl_3 \cdot NH_3$ decomposition	0.1	20	0.067
$AlCl_3 \cdot NH_3$ decomposition	2.0	20	0.85
$AlCl_3 \cdot NH_3$ decomposition	0.1	60	4.0
$AlCl_3 \cdot NH_3$ decomposition	2.0	60	23.0
Carbothermal	0.1	20	0.08

4.1.4 Boron Carbide Ceramics

Pure boron carbide is insoluble in acids like HCl, H_2SO_4 and HNO_3, even during longer boiling.

The studies on the electrochemical behaviour of hot-pressed boron carbide in H_2SO_4 and NaOH solutions [4.13] have demonstrated that the corrosion resistance of B_4C dependends to a large extent on the content of additives and impurities.

However, during heating strong oxidizing acids and mixtures such as $CrO_3 + H_2SO_4$, $KIO_3 + H_3PO_4$, $K_2Cr_2O_7 + KIO_3$, $CrO_3 + H_2SO_4 + H_3PO_4$, $KMnO_4 + H_2SO_4$, $H_2SO_4 + HClO_4 + K_2Cr_2O_7$ and others [4.9] actively oxidize free carbon present in many boron carbide materials. To remove free carbon from B_4C, alkaline solutions of hydrogen peroxide, bromine and others can also be used [4.9].

In [4.2] a sulfuric acid solution was used to investigate the resistance of hot-pressed boron carbide produced from the powder containing 77.8%–79% total boron, 21.6%–20.4% carbon, 0.25%–0.7% boron oxide, 0.25%–0.7% free boron, 0.02%–0.07% sulfur as well as Fe, Ca, Si, Al, Na, K, Mg, Ti, Ni, Cu, Cr, Pb, Sn impurities, with their total content not exceeding 0.2%. The additive-free B_4C specimens with a porosity of 1% did not interact with 10% sulfuric acid either at room temperature or on boiling. A small mass loss ($< 0.1\,\mathrm{mg/cm^2}$) in the initial period of boiling can be explained by surface impurities.

4.1.5 Boron Nitride Ceramics

As is shown in [4.9] the corrosion resistance of boron carbide powder in water, mineral acids and alkalies is largely dependent on its purity and the ordering of the crystalline structure. High-purity boron nitride powder with disordered structure, just as commercially pure, ordered BN, hydrolyses easily in humid air and decomposes completely in boiling water. Pyrolitic boron nitride with a density of $2040\,\mathrm{kg/m^3}$ and more does virtually not hydrolyse in boiling water. The lower the material density, the lower its resistance will be.

The resistance of pure boron nitride to attack by diluted acids is higher than to attack by concentrated ones and decreases in the order $HNO_3 - HCl - H_2SO_4$. Commercially pure boron nitride decomposes in diluted hydrochloric acid at a higher rate than in the concentrated one, which is apparently due to more active dissolution of interaction products in water. It interacts slowly with concentrated sulfuric acid to form ammonium sulfate and boric acid [4.1]. Disordered BN decomposes easily in acids. The resistance of pyrolytic boron nitride to acid attack depends on the specimen density, with the decomposition rate of dense specimens being extremely low. Boron nitride dissolves completely on interaction with concentrated hydrofluoric acid by the reaction $BN + 4HF = NH_4BF_4$ and on heating with ammonium fluoride and concentrated sulfuric acid by the reaction $BN + 3NH_4F + 2H_2SO_4 = 2(NH_4)_2SO_4 + BF_3$.

Boron nitride specimens are resistant enough to attack by hydrochloric, sulfuric and phosphoric acid solutions with additions of oxidizing agents ($KMnO_4$, $K_2Cr_2O_7$, $KClO_4$). They decompose most rapidly in a 5% H_2SO_4 solution with $KClO_4$ addition [4.1].

The corrosion resistance of powders and compact specimens in alkaline solutions is rather high and almost independent of the concentration.

4.1.6 Oxide Ceramics

Alumina is by far the most important oxide ceramic and the most versatile in resisting corrosion [4.17]. In the pure form, alumina has a good resistance to attack by acids and a fairly good resistance to alkalies.

With impure materials, the inertness depends on the composition of the minor phase or phases, as well as on the processing method and the grain structure achieved. Some of these materials have been studied at the National Physical Laboratory [4.18] with regard to acid attack. The test results demonstrate that, even with only 0.5% of components, resistance to attack is determined by the precise composition of the minor grain boundary phase.

For applications involving exposure to acids, some suppliers make small additions of oxides such as MgO and SiO$_2$ so that the bonding glassy phase produced during firing has a greater resistance to intergranular attack than the general purpose materials [4.18]. At the same time, it is shown in [4.19] that a higher SiO$_2$ content in alumina ceramics somewhat decreases the resistance to attack by hydrochloric acid at 80°–250° C because of dissolution of the silicate grain boundary phase.

Zirconia is one of the thermodynamically most stable ceramic materials and is appreciably more refractory than alumina. Since pure zirconia undergoes phase transformations on cooling from the firing temperature, it is necessary to include a small proportion of CaO, MgO or Y$_2$O$_3$ to stabilize the cubic or tetragonal structure. Out of these, CaO is most widely used. But the presence of a base does make the ceramics somewhat more vulnerable to acid attack [4.18].

This section is concluded with Table 4.3 summarizing comparative data on the corrosion resistance of different ceramics and providing a guidance for the choice of the material which is best suited for operation in a given medium.

Table 4.3. Resistance to acids and alkalies [4.18]

Reagent	Oxides		Nonoxides			
	Al$_2$O$_3$	ZrO$_2$	Self-bonded SiC	RBSN	HPSN	Hot-pressed BN
Acids:						
General inorganic	–	B	A	A	A	B
HCl	A	A	A 200	A	A	–
HNO$_3$	A	B	A 226	A	A	B
H$_2$SO$_4$	A	B	A 200	A	A	B
H$_3$PO$_4$	A	B	A 226	A	B	B
HF	A	B	A	B	C 20	B
Alkalies:						
General	–	–	A	B	–	A
Hot alkaline solutions	B	–	B	B	–	A
Hot KOH solution	A 80	A 80	B 80	A 80	A 80	A

A – Resistant to attack up to temperature (° C) indicated (often not given in technical literature)
B – Some reaction at temperature indicated
C – Appreciable attack at temperature indicated

4.2 Molten Salt and Alkali Corrosion

The studies on molten salt corrosion of structural ceramics are of interest, since these materials are used for manufacturing components operating in aggressive environments, e.g. electrolytic baths, gas-turbine engines and heat exchangers in which the firing of low-grade fuel results in Na_2SO_4, V_2O_5 and other compounds. Marine-based turbine engines can also inject sea salt spray.

In [4.20] corrosion of hot-pressed and reaction-bonded silicon nitride and silicon carbide (Norton Co.) was investigated in molten Na_2SO_4, NaCl and eutectic Na_2SO_4–NaCl (63 mole % Na_2SO_4 and 37 mole % NaCl). The results are presented in Fig. 4.5. The materials are most actively attacked by molten sodium sulfate, with silicon nitride being more stable under these conditions than silicon carbide. A certain mass gain of the specimens after corrosion in molten NaCl (Fig. 4.5) is determined by their oxidation with oxygen of the air dissolved in the molten salt or by the diffusion implantation of sodium and chlorine in the structure of the material.

In [4.21] corrosion of HPSN and RBSN as well as self-bonded SiC was studied in molten V_2O_5 and the equimolar mixture Na_2SO_4–V_2O_5. In the latter case the reaction $Na_2SO_4 + V_2O_5 = 2\,NaVO_3 + SO_3$ was assumed to proceed in the melt. As the experiments have demonstrated, the exposure to molten V_2O_5 at 900° C for 240 h does not exert a considerable influence on the materials. In contrast to this, the equimolar mixture affects the material in almost the same way as pure sodium sulfate. And the resistance of HPSN is about an order higher than that of RBSN. Preoxidation in air does not produce a protective effect on the materials, since sodium sulfate dissolves SiO_2 in accordance with reaction (3.1) and the following reaction:

$$Na_2SO_4 + 2\,SiO_2 = Na_2Si_2O_5 + SO_3 \quad . \tag{4.1}$$

In [4.21] the Gibbs energy of reaction (4.1) was determined at different temperatures (Table 4.4.).

Table 4.4 Gibbs energy (ΔG) of reaction (4.1) [4.21]

T, °C	827	927	1027	1127	1227
ΔG, kJ/mole	159.6	142.8	126.0	109.2	96.6

As follows from these data, the probability of this reaction increases with temperature.

RBSN with a porosity of 30% is resistant to the molten eutectics LiCl–KCl and LiF–LiCl–KCl at 400° C. After exposing the specimens to these melts for 1000 h their mass loss was rather small and any noticeable changes of the surface structure did not occur [4.22].

In the molten NaF and ZrF_4 mixture silicon nitride decomposes at 1100° C in 4 h, while at 900° C it does not react for 144 h [4.9]. Silicon nitride does not

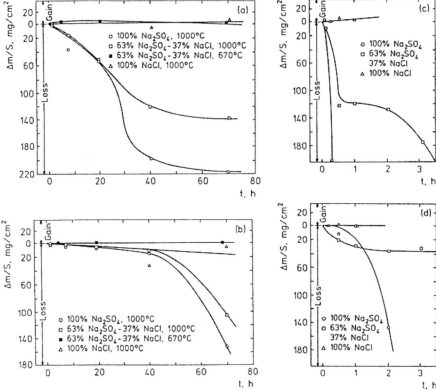

Fig. 4.5. Mass change vs time of exposure to molten salts of RBSN (a), HPSN (b), self-bonded SiC at 1000° C (c) and HP SiC at 1000° C (d) [4.20]

interact with molten lithium and lanthanum fluorides up to 500° C and with molten $NaNO_3$–$NaNO_2$ at 350° C.

Reaction-bonded ceramics of the system Si_3N_4–SiC of an NKKKM–81 grade were also examined for their corrosion behaviour. As the investigations have demonstrated [4.23], these ceramics degrade rapidly in molten Na_2SO_4. Their exposure to this melt at 1050° C for 1.5 h results in almost complete dissolution of the specimen (Fig. 4.6). At 950° C NKKKM–81 is slightly more stable in molten sodium sulfate; however, after exposure for 1 h (Fig. 4.7a) the surface

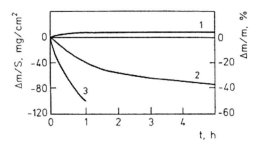

Fig. 4.6. Corrosion kinetics of NKKKM–81 ceramics in molten NaCl and sea salt at 1050° C (1) and in molten Na_2SO_4 at 950° (2) and 1050° C (3)

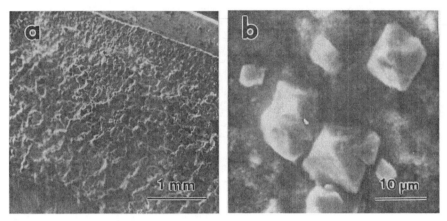

Fig. 4.7. Surface areas of the specimens exposed to molten Na_2SO_4 at 950° C for 1 h (a) and to molten sea salt at 1050° C for 1.5 h (b)

layers of the specimens contain only β-Si_3N_4, and the whole silicon carbide, which is less stable under these conditions, dissolves in the melt.

The corrosion processes of NKKKM ceramics in sodium sulfate can be described in the following way: They start oxidizing at temperatures close to 800° C, and salt melting occurs only at 884° C. Therefore, on heating in the temperature range of 800°–900° C the surface layer of ceramics is oxidized with oxygen of the air according to reactions (2.1, 14).

After melting the salt starts to interact with the silica layer by reactions (3.1) or (4.1) leading to its dissolution. The dissolution of the oxide layer initiates the reaction between the salt and Si_3N_4 and SiC:

$$6Na_2SO_4 + 2Si_3N_4 + 3O_2 = 6Na_2SiO_3 + 4N_2 + 6SO_2 \qquad (4.2)$$

and

$$Na_2SO_4 + SiC + O_2 = Na_2SiO_3 + CO + SO_2 \quad . \qquad (4.3)$$

Reaction (4.3) proceeds rather actively. The SO_2 evolution was revealed by a specific odour and by wet litmus paper turning red. It is also interesting to note that a chemical analysis of the dissolved cake detected SO_3^{2-} ions. A certain fraction of SO_2 apparently dissolves in molten salt and passes into the solution after the dissolution of the cake in water.

The process takes place in this way if there is free access for air, with an open crucible and a molten layer of several mm over the specimen. In a tightly closed crucible, without access for air oxygen, the corrosion rate decreases 2–3 times.

As opposed to Na_2SO_4, the molten sea salt of the Black Sea basin with the following composition, %: 1.402 NaCl, 0.0189 KCl, 0.1304 $MgCl_2$, 0.0005 $MgBr_2$, 0.0105 $CaSO_4$, 0.047 $MgSO_4$, 0.0359 $Ca(HCO_3)_2$, 0.0209 $Mg(HCO_3)_2$, and its major component NaCl induce a small mass variation of NKKKM specimens (Fig. 4.6). The mass gain can be explained by the oxidation of the specimens with the oxygen of the air until salt melting or with oxygen dissolved in the molten salt as well as by the diffusion implantation of sodium, chlorine and other atoms in the structure of the ceramics and by an incomplete removal of residual salt during rinsing.

The studies on the interaction of NKKKM–81 with molten sea salt revealed octahedral crystals of cubic syngony (Fig. 4.7,b). These crystals grow in the direction $\langle 100 \rangle$. EDAX detected magnesium and silicon in their composition. The only compound which can develop under the given conditions is magnesium silicate. However, oxygen could not be determined, since chemical elements lighter than sodium were not detectable by the EDAX analyser. Magnesium silicate can form on the interaction of magnesium ions present in the molten salt with SO_3^{2-} ions formed as a result of reactions (4.2–5). At the same time magnesium is present in the ceramics as a component of the secondary grain-boundary phase. The formation of magnesium silicate on the surface of crystals can involve the transport of magnesium ions through the liquid phase. To elucidate the mechanism of crystal formation, the specimens exposed to molten NaCl were thoroughly investigated by optical and electron microscopy. No traces of such crystals were revealed on them. This confirms that the magnesium ions stem from the molten sea salt. It is interesting to note that enstatite and forsterite crystals refer to rhombic syngony, while the crystals obtained under the conditions of our experiment were regular octahedrons.

Molten alkalies decompose silicon nitride with ammonia evolution. This phenomenon is used, e.g., for the etching of polished samples in ceramography. Molten sodium peroxide and alkali carbonates also interact with Si_3N_4. Lead chromate, lead dioxide and monoxide decompose it with nitrogen evolution. A mixture of one part of $PbCrO_4$ and one part of PbO and PbO_2 starts reacting actively with Si_3N_4 at 500° C. This reaction is used for determing the nitrogen content in silicon nitride [4.1].

Silicon carbide also degrades quite easily in molten alkalies, basic salts, alkali carbonates: $SiC + Na_2CO_3 + 2O_2 = Na_2SiO_3 + 2CO_2$ [4.24]. The data on the resistance of SiC single crystals in different melts are summarized in Table 4.5.

Table 4.5. SiC resistance to melts [4.25]

Corrosion environment	T_m/T_{etch}, ° C	V_{etch}, μm/min
Na_2CO_3	851/900	0.3
$Na_2B_4O_7$	741/1000	0.3
NaOH	318/900	1.5
Na_2O_2	314decomp./700	7.0
$Na_2O_2 + NaOH$ (1 : 3)	–/700	7.0
$NaF + Na_2SO_4$ (1 : 1)	980, 884/950	5.0

T_m – melting point
T_{etch} – test temperature
V_{etch} – etching rate

Sintered α-SiC corrodes in molten $Li_2SO_4|Li|LiF$, while molten $Li_2S|Li|LiF$ and $LiCl|Li|LiF$ affect silicon carbide ceramics to a lesser extent [4.26].

Sintered aluminium nitride with a porosity of 25%, just as silicon nitride, is resistant to the molten eutectics LiCl–KCl and LiF–LiCl–KCl at 400° C [4.22] but degrades easily in molten alkalies. Aluminium nitride also interacts easily with lead dichromate. Sodium peroxide decomposes it with the formation of nitrates.

Table 4.6. Resistance to fused salts, alkalis and low-melting oxides [4.28]

Medium	Silicon nitride		Silicon carbide	Boron nitride, pyrolytic	Alumina	Zirconia, stabilized
	RBSN	HPSN				
Acid slags					A	A
Basic slags					B	A
Na_2CO_3	C550 (Air)	C900 (Air)	C900 (Air)	C900 (Air)	A900	C900
KNO_3	A400		A400	A400	A400	
Na_2SO_4	C1000 (Air)	B1000 (Air)	C1000 (Air)	C1000 (Air)	A1000	A1000
K_2SO_4			C		A1140	A1140
$KHSO_4$	A500		A500	C500	C500	C500
LiCl			B	A	A620	
NaCl		A	A900 (Air)		A	
NaCl+KCl	A790		C800		A800	C800
$BaCl_2$	C1000 (Air)		C1000 (Air)		B1350	B1000
Na_3AlF_6			C	A1050	C	
KF	C900 (Air)		C900 (Air)	A1060	C900	B900
$Na_2B_4O_7$	A1000	B	B1000	B1000	C1000	C1000
$NaVO_3$		C800 (Air)	C800 (Air)		B800	A800
NaOH	C450	C500	C500		A538	A538
KOH	C500	C500	C500	C500	B500	B500
Na_2O_2	C500	B	C750	C500	B500	C500
PbO	C1000 (Air)	C1000 (Air)	C1000 (Air)	C1000 (Air)	B1000	A1000
V_2O_5	C800	C800	C800		C800	C800

Key:
A - resistant to attack up to temperature indicated (°C),
B - Some reaction at temperature indicated,
C - Appreciable attack at temperature indicated
(Air) - Indicates atmosphere used for experiment.

Boron nitride fused with potassium carbonate decomposes by the reaction $BN + K_2CO_3 = KBO_2 + KOCN$. Nitride excess leads to a certain amount of KCN. If carbon is present, the reaction proceeds as follows:

$$4BN + 3K_2CO_3 + 2C = K_2B_4O_7 + CO_2 + 4KCN \quad .$$

Boron nitride reacts with sodium formate [4.1]. At the same time it is necessary to note that boron nitride has an exceptionally good resistance to molten salts and glasses. This is probably due to its resistance to wetting. It is inert to a number of corrosive molten salts, e.g. alkali halides, lithium borate and cryolites. However, it does react with molten alkali carbonates and hydroxides. Hot-pressed boron carbide is attacked by fused alkalies.

High-alumina ceramics are probably the most resistant oxide material, but even they are heavily attacked by melts of Na_2O, $Na_2S_2O_7$, Na_2O_2 and KOH. Other ceramics may be preferable for particular melts, e.g. spinel is reported to be more resistant to basic slags, while zirconia is preferred for melting glasses containing phosphates and borates [4.18].

Fused sodium carbonate attacks most ceramics to some extent, but aluminas are among the most resistant.

The corrosion of 3 mol% Y–PSZ with alkaline metal fluorides depends on the ionic radius of univalent cations. The activation energies to estimate the $t \rightarrow m$ transformation rate of 3Y–PSZ, 3Y–PSZ–Al_2O_3 and 3Y–PSZ–Cr_2O_3 are 12, 44 and 154 kJ/mole, respectively. The phase stability of 3Y–PSZ–Al_2O_3–Cr_2O_3 prepared by an impregnation method is the highest in the molten fluoride salts [4.27].

This section is concluded with Table 4.6 which summarizes comparative data on corrosion resistance of different ceramics and helps to choose the material most suitable for operation in a given medium.

5. Corrosion Effect on Ceramic Properties

The knowledge of the corrosion behaviour of ceramic materials is rather important for determining their fields of application (refractories, high-temperature insulation and others), especially when oxidation-induced variations of mechanical properties (including local damages) have no effect on the performance of a ceramic design. However, to specify the load carrying capacity of components of gas turbines and Diesel engines, this information is insufficient, since corrosion-induced variations of composition and structure of a ceramic surface influence the strength of the material (Table 5.1), lead to the failure of components and the engine itself during operation. Oxidation also exerts a considerable influence on the physical properties, e.g. thermal and electrical conductivity. It involves the growth of SiO_2 contents in ceramics and an increase of their thermal linear expansion coefficients as a result [5.1]. But conclusions of possible oxidation-induced changes of ceramic properties can be made only from the investigations of corrosion mechanism and kinetics. Therefore, the impact of corrosion on the physico-mechanical properties of ceramics requires further examination.

5.1 Preoxidation of Ceramics

The oxidation of ceramics in air can exert a great influence on their strength at room temperature (Table 5.1). The results may be explained on the basis of the data on composition and structural changes of the surface layer after oxidation of ceramics cited in Chap. 2. Thus, it is quite evident that cracking of specimens on oxidation (Fig. 2.49) will cause a reduction in strength, just as pitting or other flaws. At the same time the factors which are responsible for higher strength after oxidation (Table 5.1) are not so obvious. Also, visual inspection alone cannot reveal the cause of strength change of ceramic specimens after heating in oxidizing environments. Therefore it is quite appropriate to analyse in more detail the effect of oxidation on the properties of different ceramics.

5.1.1 Silicon Nitride Ceramics

We should elucidate how strength is influenced by oxidation time, cooling rate of specimens, state of their surface etc., to get information which is applicable in practice for the development of advanced ceramics. During experiments

Table 5.1. Strength changes and oxidation mass gain for commercial Si_3N_4 and SiC due to static air exposure for 1000 h [5.2]

Material	Exposure temp.,	Strength change/25° C,	Fracture origins		Mass gain,
	° C	%	Unexposed	Exposed	%
Si_3N_4 Materials					
HPSN (1% MgO)	1200	−33	Inclusions	Oxidation pits	0.48
(4% MgO)	1200	−41	Inclusions	Oxidation pits	2.6
HPSN (8% Y_2O_3)	1000	+7	Inclusions	Inclusions	0.05–0.10
HPSN (15% Y_2O_3)	1000	+10 −100	Inclusions	Internal cracking due to accelerated oxidation	0.03–5.1
HPSN (4% Y_2O_3, 3% Al_2O_3)	1000	No change	Inclusions	Inclusions	0.06–0.58
HPSN (4% Y_2O_3, SiO_2)	1000	+10	Inclusions	Inclusions	0.01
HPSN (10% CeO_2)	1000	−50	Inclusions	Oxidation pits	0.32
RBSN	1400	+3 to −60	Inclusions, porosity	Primarily oxidation pits	–
SiC Materials					
HPSiC (2% Al_2O_3)	1400 1500*	−10 0	Undetermined	Undetermined	0.1–0.2
HPSiC (2% B_4C)	1400	−4	Undetermined	Oxidation pits	0.49
Sintered α-SiC	1400	−8	Inclusions, grains, pores	Oxidation pits	0.08
Si/SiC	1200	+30 to −46	Undetermined	Mostly undetermined, some oxidation pits	0.06–0.16

* 2000 h

with reaction-bonded SiC-containing NKKKM–79 ceramics (see Sect. 2.1.4 and [5.1]), an increase of oxidation time over 3 h did not cause any further changes of strength (Fig 5.1a). This corresponds to the data of [5.3] obtained for RBSN tested for 100 h. Thus, to simplify the experimental procedure for studying the effect of the oxidation temperature on the strength of ceramics, the exposure of specimens was restricted to 3 h.

The studies on the effect of the cooling rate (Fig. 5.1b) have demonstrated that the quenching of specimens in air from 900° C does not change their strength at 20° C, the quenching from 1050° C leads to its increase and the

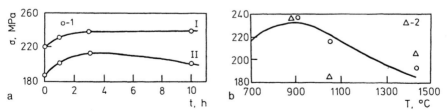

Fig. 5.1. Strength as a function of exposure time at 900° C (a) and oxidation temperature (b) for polished NKKKM–79 specimens: test temperature 20° C (I), 1400° C (II); quenching (1), cooling with a furnace (2)

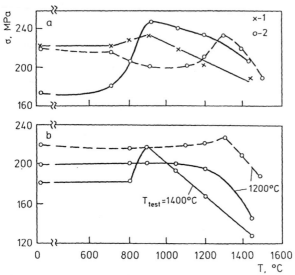

Fig. 5.2. Strength as a function of oxidation temperature for polished (1) and unpolished (2) NKKKM–79 (solid lines) and NKKKM–83 (dashed lines) specimens at room (a) and higher (b) temperatures

quenching from 1400° C results in its decrease. Since the temperatures below 1400° C are of interest for practical applications, the specimens after oxidation were removed from the furnace and cooled in air.

Figure 5.2 shows that the strength of as-received specimens after oxidation at room temperature becomes higher than that of polished specimens. The healing of flaws with oxidation products not only increases strength but also reduced data scattering [5.4]. This is observed even if the specimens are oxidized during their heating for strength tests (Fig. 5.3).

It is characteristic that oxidation leads to a higher strength than polishing of specimens, since defects are also healed in the subsurface layer of the ceramics. This phenomenon should be taken into account in further investigations of the effect of oxidation on the strength of ceramics and in the evaluation of strength tests of ceramic specimens. The test results for polished specimens are often used to estimate the strength of components which operate without any kind of surface treatment (as-sintered).

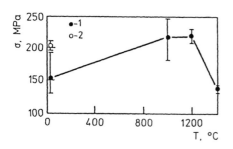

Fig. 5.3. Strength as a function of temperature for as-received (1) and oxidized at 1200° C (2) NKKKM–83 specimens (1.5 mm thick)

Since the strength of as-sintered specimens after oxidation becomes higher than that of the polished ones and the components produced from these ceramics are not machined, all the experiments to study the effect of oxidation on strength were performed with as-sintered ceramic specimens.

With an increase in oxidation temperature (Fig. 5.2a), the strength of NKKKM–79 specimens at 20° C starts decreasing because of pores, blisters and other flaws appearing in the oxide layer or because of the etching of grain boundaries.

For more resistant NKKKM–83 ceramics the effect of oxidation on strength at room temperature is not so pronounced as for NKKKM–79 ceramics (Fig. 5.2a). An increase in the strength of specimens occurs at 1300°–1350° C and makes up only 15% (40%–45% for NKKKM–79). To corroborate the fact that higher strength of specimens after oxidation at room temperature is caused by the healing of flaws, specimens with a thickness of 1.5 mm were tested. The thinner the specimen, the greater should be the influence of surface flaws (Figs. 5.2a and 5.3). The observed difference in strength of 30% confirms the fact that the thickness of a defective surface layer probably becomes comparable with the thickness of the specimen. At the same time the strength values of 1.5 mm thick specimens at 1200° C were almost the same as those of 3.5 mm thick specimens, since flaws are healed by oxidation occurring on heating of specimens up to test temperature and 15 min exposure at a given temperature. To demonstrate that a higher strength at 1000°–1200° C is due to oxidation and not to any other factors, the specimens were heated up to 1200° C, then cooled and tested at room temperature (Fig. 5.3). In this case the strength values corresponded to the specimens of standard sizes (Fig. 5.2a).

We should note that oxidation at 1400° C and above leads to a decrease of strength of both NKKKM–83 and NKKKM–79 ceramics (Fig. 5.2a). A certain reduction of NKKKM–83 strength after oxidation at 800°–1100° C, also typical of many other materials [5.5], can be caused by cracks formed in the oxide layer, which is usually brittle as these temperatures (see also Sect. 2.1.6).

Oxidation influences the strength of ceramics not only at room temperature but also at higher temperatures (Fig. 5.2b). Thus, the strength of both ceramic modifications at 1200° C after oxidation up to this temperature remains almost the same, but after oxidation at 1400° C and above decreases considerably. NKKKM–79 ceramics behave similarly at 1400° C and 20° C (Fig. 5.2a).

The investigations demonstrated how the number of defects in the surface layer affected the strength of the ceramics. During oxidation the surface layer of the specimens also concentrates impurities, i.e. the inner layers of the material are cleared from them (Sect. 2.1). Therefore, an investigation was carried out to elucidate the effect of an oxidized surface layer on the strength of ceramics.

For this investigation we used NKKKM–83[1] specimens (Table 5.2) oxidized at 1400° C for 5 h. During this time a 0.7 mm thick oxidized layer consisting of Si_3N_4, SiC and their oxidation products appears on their surface. Oxidized specimens were ground to remove the surface layer as completely as possible

---

[1] For these investigations NKKKM–83 from another lot was used.

Table 5.2. Test results for differently treated specimens

Group of specimens	Treatment	Phase composition of treatment specimens	σ, MPa
1	Grinding	β-Si_3N_4, SiC, α-Si_3N_4 and Fe_5Si_3 traces	252.1/120.0
2	Oxidation + grinding	β-Si_3N_4, SiC, Fe_5Si_3 traces	299.0/94.7
3	Grinding + oxidation	α-cristobalite, SiC, β-Si_3N_4, (Ca, Mg, Fe)SiO_3, tridymite, $MgSiO_3$ and Mg_2SiO_4 traces	231.2/78.5
4 .	Oxidation + grinding + oxidation	α-Cristobalite, SiC, β-Si_3N_4, tridymite, (Ca, Mg, Fe)SiO_3	236.4/97.1

Note. The values of σ at 20° and 1400° C are given in numerator and denominator, respectively

and to exclude its influence on the test results. The XRD analysis did not reveal any crystalline oxide phases in the composition of the specimens of group 2.

Crystalline Mg, Ca, Fe compounds (Table 5.2) in the surface layer of the specimens of groups 3 and 4 were not detectable by XRD and made up 5%–10%, i.e. the impurities with high affinity for oxygen concentrated in a thin film on the specimen surface. The crystallization of impurity phases in the oxide layer was achieved by slow cooling of specimens in the furnace. On rapid cooling of specimens designed for strength tests, impurities become a part of the glassy phase.

The specimens which were not exposed to preoxidation exhibited the highest strength at 1400° C (Table 5.2). However, the strength of the specimens of groups 2 and 4 is 15%–20% higher than that of the specimens of group 3, since due to preoxidation and removal of an oxide layer they contained a smaller amount of impurities. Thus, the content of liquid phase formed on their oxidation was lower and it was of higher viscosity. A lower content of impurities in the specimens is confirmed by EDAX data.

A slight increase in high-temperature strength after the removal of a surface layer is apparently due to the fact that the porous materials under study are oxidized not only on the surface but also across the thickness of the specimen. Thus, the oxidation has a stronger adverse effect on porous materials than on dense ones [5.6].

Oxidation at 1400° C slightly decreases the strength of the specimens of groups 3 and 4 at room temperature (Table 5.2 and Fig. 5.2a) because of pores and other flaws formed in a solidified oxide layer. The surface of several specimens after oxidation and quenching in air revealed a network of small cracks. At the same time the strength of the specimens of group 2, with the oxide layer removed, became 20% higher compared to the specimens of group 1. This increase in strength is probably due to the healing of defects in the inner layers of the material. Since XRD could not detect an oxide phase in

Table 5.3. Properties of NKKKM-85 specimens cut from components

Component	1	2
σ, MPa, at T, ° C:		
20	168	168
1200	148	127
K_{1c}, MPa $\cdot$ m$^{1/2}$, at T, ° C:		
20	2.50	2.74
1200	2.71	2.23

these specimens, one may suppose that oxidation results in a thin layer of amorphous silica formed on the surface of open pores.

The cracking of an oxide layer as a result of oxidation can be caused by stresses arising due to the increase in volume on the conversion of Si_3N_4 and SiC into SiO_2 or by phase transformation in cristobalite above 200° C changing the volume by about 5%.

The reduction in strength after oxidation can be caused by the cracking of a surface layer, the etching of grain boundaries, the formation of a liquid phase along the grain boundaries as well as pores, craters and other flaws in the oxidized surface layer of ceramics.

The obtained results were used for developing recommendations on the preoxidation of different NKKKM modifications to improve their performance characteristics. The data on NKKKM-85 properties determined on the specimens cut from ceramic components are given in Table 5.3.

Component 1 was oxidized in air under preset conditions, and component 2 was investigated as-received. The kinetic curves of oxidation for these specimens are shown in Fig. 5.4. Preoxidation produced a favourable effect on the characteristics of the ceramics. On the specimens cut from component 1, the liquid phase was formed in smaller quantities and at higher tamperatures than on the specimens cut from component 2. The oxide layer formed on the specimens from component 1 contained much less blisters and pores.

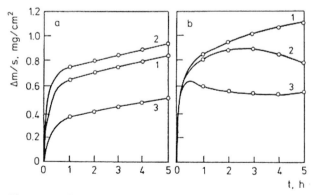

Fig. 5.4. Oxidation kinetics of NKKKM-85 specimens cut from components 1(a) and 2(b) at 1200°(1), 1300°(2) and 1400° C (3)

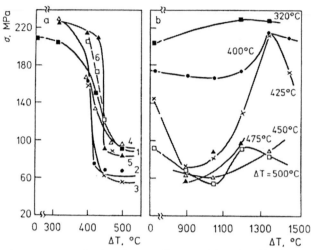

Fig. 5.5. Termal shock resistance diagrams (a) and residual strength as a function of oxidation temperature (b) for NKKKM–83 ceramics: as-received specimens (1), after oxidation at 900°(2), 1050°(3), 1250°(4), 1350°(5) and 1450° C (6)

We also investigated the effect of oxidation on the thermal shock resistance of NKKKM–83 ceramics [5.7]. Oxidation leads to rather critical changes of the residual (after quenching) strength of ceramics (Fig. 5.5); however, the critical quench temperature difference (ΔT_c) varies by no more than 20°–25° C. The highest increase in ΔT_c is observed after oxidation at 1350° C, i.e. when the increase in the strength of these ceramics is at its maximum (Fig. 5.2). At the same time oxidation at 900°–1050° C sharply decreases the residual strength of specimens after quenching with a temperature difference $\Delta T > \Delta T_c$. On the whole, oxidation exerts a greater influence on the slope of thermal shock resistance diagrams and residual strength of ceramics than on the value of a critical temperature difference.

The character of strength changes for NKKKM ceramics after oxidation was similar to that revealed for reaction-bonded sialons [5.8] and NR 115 RBSN (AnnaWerk) at temperatures up to 1400° C [5.3]. Oxidation in air in the range of 1000°–1400° C was reported to increase not only the critical stress intensity factor and effective surface energy but also the elastic modulus of reaction-bonded ceramics. This is apparently due to a simple increase in their density.

Oxidation of low-strength RBSN with large pores at 950°–1000° C [5.9] involved a 45% increase in strength at room temperature [just as NKKKM–79 (Fig. 5.2)] and a 65% increase at 1500° C, whereas high-strength pure RBSN [5.10] did not display any changes of strength and fracture toughness as a result of oxidation. The authors of [5.11] report a lower bending strength caused by oxidation, and this strength decreases with exposure time. Such contradictory data in the literature can be explained by the fact that oxidation of various materials leads to different changes of their structure (Chap. 2) and to a different effect on the mechanical properties.

96

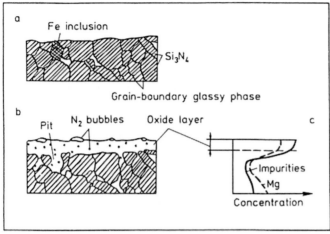

Fig. 5.6. Scheme of the surface layer of a specimen as-received (a), after heating in air above 1100° C (b) and distribution of cations in the oxidized layer (c)

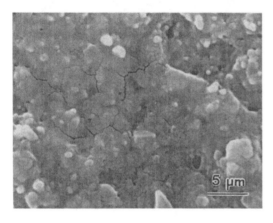

Fig. 5.7. Electron micrograph of the fracture surface of a ceramic specimen with 5% Y_2O_3 and 2% Al_2O_3 after tests in air at 800° C

Oxidation of dense SSN and HPSN degrades their strength much more often than it improves it. The strength degradation of dense Si_3N_4 may be due to the etching of grain boundaries [5.12] (Fig. 5.6), the formation of pits and pores at the oxide layer/material interface [5.13] or the cracking of the surface layer due to oxidation of the grain boundary phase [5.5] (Fig. 5.7). The observed variations of properties are determined by the composition and structure of the ceramics.

Thus, the manufacturing process and oxidation conditions can result in different strength changes after the exposure of the ceramics to air or oxygen at higher temperatures (Table 5.1). During oxidation of ZrO_2-doped hot-pressed materials, compression stresses in the surface layer appear already at 600°–800° C [5.14]. This is caused by the presence of the secondary phase consisting of $Zr_{2-2x}N_{4x/3}(0.25 \leq x \leq 0.43)$ which is oxidized to form monoclinic ZrO_2

above 500° C; in turn, the volume increases by 45%. At the temperatures of the process either Si_3N_4 or Si_2N_2O has not yet been oxidized. Due to high density, the secondary phase oxidizes only at the surface. Compression stresses improve strength by hindering crack initiation in the surface layer of the material. However, when oxidation is too active, the oxide layer cracks because these stresses exceed a permissible level. A similar picture was also observed for CeO_2-doped materials.

Oxidation of MgO- and SiO_2-doped HPSN for 300 h at 1400° C followed by grinding-off the oxide layer removes impurities, which leads to a 1.5 fold increase in strength at 1400° C. The value of K_{1c} grows from 6 to 8.5 MPa · m$^{1/2}$ already after 5 h of oxidation [5.6].

Oxidation at 1375° C for 375 h followed by grinding of the surface extends the time to rupture under load of NC–132 MgO-doped HPSN (Norton Co.) [5.15]. But oxidation at 1450° C abruptly decreases long-term strength of the material.

The bending strength of SSN and HPSN was investigated after oxidation and oxidation under load at 1370° C [5.16]. Oxidation was shown to reduce the strength of the materials. Oxidation under load increased the SSN strength and decreased that of HPSN due to the growth of nitride grains.

The above data referred to strength values obtained during the bending tests. Oxidation-induced changes of specimen surfaces are shown to exert a greater influence on bending than on tensile strength due to a nonuniform distribution of stresses [5.17].

According to the data of [5.18], oxidation of Y_2O_3-doped HPSN for 120 h at 1600° C considerably decreases the creep rate at 1400° C. This is apparently due to the removal of impurities from the inner layers of the material, a lower grain-boundary sliding rate and smaller diffusional creep.

According to the data of [5.12], longer oxidation times (over 200 h) do not involve a further strength degradation of HPSN (Fig. 5.8), probably because longer exposure times do not enlarge flaw sizes (pits and grooves along the grain boundaries). Lower strength of HPSN and sialons can also be caused by oxidation of free silicon interlayers [5.19].

Significant increases in the strength of HIPSN were observed after exposure for 10 h at 1400° C in atmospheres of H_2 containing relatively large amounts

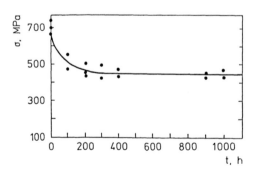

Fig. 5.8. Strength degradation of MgO-doped HPSN as a function of heating time at 1200° C [5.12]

of H_2O ($P_{H_2O} \geq 1 \cdot 10^{-4}$ MPa) and Ar with intermediate amounts of O_2 ($7 \cdot 10^{-6}$ MPa $\leq P_{H_2O} \leq 1.5 \cdot 10^{-5}$ MPa) [5.20]. Healing of surface flaws by the formation of a $Y_2Si_2O_7$ layer was assumed to be responsible for this increase in strength. When the P_{H_2O} in H_2 was equal to $2 \cdot 10^{-5}$ MPa, the material loss was high (as a result of an active oxidation process) and the bending strength decreased. When the P_{O_2} in Ar was equal to $1 \cdot 10^{-3}$ MPa, SiO_2 was formed on the surface in addition to $Y_2Si_2O_7$, and the strength was similar to that of the unexposed material. Under this condition, the previously observed beneficial crack healing effect was apparently nullified by the generation of new flaws, such as cracks and blisters, in the oxide layer.

On the basis of the above-mentioned results we can formulate the causes of the strength loss by the material after oxidation:

- cracking of the surface layer caused by phase transformation in cristobalite (this often causes a drop of strength at room temperature after oxidation at 1100° C or more; it should also be taken into account that the level of this drop is determined by the depth of the arising cracks which, in turn, depends on the thickness of the oxide film);
- the compression stresses induced in the surface layer by the difference in the thermal expansion coefficients of the material and of the oxide layer, and by the volume increase in the process of conversion of Si_3N_4 and SiC into SiO_2 (all this may also increase the strength; however, when the stresses are very high, the oxide layer is subject to cracking and the strength decreases [5.21]);
- stresses arising on heating because of the difference in the thermal expansion coefficients of the compounds Si_3N_4, SiC and the intergranular phase observed at temperatures of 600°–900° C which somewhat impairs the strength of the material;
- etching of the grain boundaries and formation of a liquid phase along them in the surface layer (this is one of the causes of reduced high-temperature strength after oxidation at temperatures above 1100° C, when a liquid oxide phase forms on the surface of the material);
- formation of oxidation pits.

The increase in strength can be caused by the healing of pores, cracks and other flaws in the surface layer as well as defects in the inner layers of the specimens, by lower contents of impurities in the ceramics due to their diffusion to the surface of the specimens during oxidation, by compression stresses arising in the surface layer due to the difference in the thermal expansion coefficients of the material and the oxide phase and due to an increase in volume on conversion of Si_3N_4 and SiC into SiO_2.

5.1.2 Silicon Carbide Ceramics

Oxidation exerts different effects on the strength of various SiC-based ceramics (Table 5.1). Thus, by the data of [5.22], at room temperature the bending strength of self-bonded SiC decreases after oxidation, while it does not change for HPSiC and increases for recrystallized SiC. At the same time, at room temperature the strength of sintered and self-bonded SiC is reported to increase after oxidation for 5000 h at 1200° C [5.23]. Similar heating of NC–203 HPSiC (Norton Co.) decreases its strength from 600 to 300 MPa. It is shown in [5.24] that oxidation at 900° C does not influence the strength of HPSiC, but oxidation for 1000 h at 1260° C under complex cyclic conditions decreases its strength from 700 to 350 MPa at 20° C. The contradictory data of [5.22] and [5.18, 23] are apparently due to different oxidation conditions (oxygen pressure, temperature, exposure time, cyclic or static heating etc.) which may result both in higher and lower strengths of different materials. Oxidation mechanisms influencing the strength of silicon carbide ceramics are similar to the ones described above for silicon nitride materials.

The author of [5.24] investigated the effect of oxidation on the strength of Crystar recrystallized SiC specimens with 20% porosity and dense HPSiC specimens (Norton Co.) which were subjected to a thermal shock when their surface was covered with cracks having a length of up to 1/4 of the specimen thickness. Oxidation in air at 1400° C for 90 h completely recovers the strength of recrystallized SiC specimens but only slightly affects the strength of HPSiC. The doubling of the Crystar strength is determined by the healing of cracks with silica formed as a result of oxidation. The specimens heated in vacuum at the same temperature did not display any increase in strength. So it was the oxidation but not diffusion processes that increased the strength of SiC. However the strength of self-bonded SiC specimens with a performed defect can increase by 70% after heating at 1400° C due to the diffusion or melting of free silicon present in their composition [5.26]. The results of mechanical tests at different loading rates for 9 commercially available SiC-based materials are described in [5.27]. Before the tests the specimens were preoxidized for 100 h at 1400° C. The obtained data were used to determine the values of N characterizing the resistance of materials to subcritical crack growth by the equation $\sigma = B\dot\sigma(1/N + 1)$, where $\dot\sigma$ is the loading rate, B is the constant. Sintered and hot-pressed materials were the least stable, their N values did not exceed 70 over the temperature range of 1000°–1400° C. For recrystallized ceramics N was over 100, and for self-bonded SiC at 1400° C its value reached 250. High resistance of self-bonded SiC to subcritical crack growth is apparently caused by crack retardation due to the plastic flow of free silicon. Initial specimens of self-bonded SiC are also more stable than those of hot-pressed SiC in which the subcritical crack growth occurs owing to the secondary phase present along the grain boundaries. Unoxidized sintered α-SiC did not display any relation between failure stresses and loading rates [5.28]. The subcritical crack growth in the oxidized material was induced by borosilicate glass formed on the surface and along the grain boundaries in the subsurface layer of the specimens. The failure of specimens under these conditions can also be considered as stress corrosion cracking (Sect. 6.3).

According to the data of [5.29], the oxidation of particulate SiC–TiB$_2$ composites in air at 1200°–1400° C has a slight effect on their mechanical properties. The studies on the effect of oxygen pressure on the strength of commercial sintered α-SiC at room temperature [5.30] demonstrated that on active oxidation

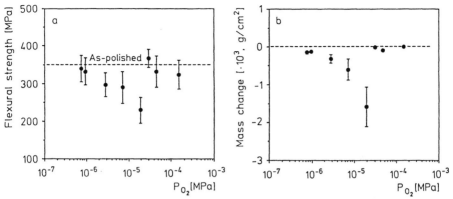

Fig. 5.9. Effect of partial oxygen pressure on the strength (a) and mass change (b) of α-SiC exposed for 10 h at 1400° C to various Ar–O$_2$ atmospheres [5.30]

the decrease of strength was proportional to the O$_2$ pressure (Fig. 5.9a) and exposure time; the mass loss had a similar trend (Fig. 5.9b).

A strength reduction of approximately 50% was observed in SiC after oxidation for 20 h at a P_{O_2} of $1.5 \cdot 10^{-5}$ MPa, a level that is slightly lower than the P_{O_2} at which the transition from active to passive oxidation occurs. Large pits formed during exposure were responsible for the reduction in strength.

When the P_{O_2} was higher than $3 \cdot 10^{-5}$ MPa, SiO$_2$ was formed on the surface (i.e. passive oxidation occurred) and the strength of the specimens was not significally affected.

5.1.3 Aluminium Nitride Ceramics

The effect of oxidation on the mechanical properties of AlN-based materials has been investigated less [5.31] than that of Si$_3$N$_4$- and SiC-based materials. However, it is known that the mechanisms of the processes leading to changes in their strength after oxidation are similar to those established for Si$_3$N$_4$ and SiC ceramics.

Oxidation in air above 800° C for 48 h decreases the bending strength of all the materials under study (Fig. 5.10). A drop of strength is apparently caused by pores and other flaws formed in the oxide layer consisting of α-Al$_2$O$_3$ on the additive-free material, of CaO $\cdot$ 2 Al$_2$O$_3$ and CaO $\cdot$ 6 Al$_2$O$_3$ on the CaO-doped material and of 3Y$_2$O$_3 \cdot$5Al$_2$O$_3$, YAlO$_3$ and AlOH on the Y$_2$O$_3$-doped material. During oxidation of the materials with oxide additives, calcium and yttrium diffuse to the surface which results in the formation of considerable amounts of calcium aluminates and yttrium-aluminium garnet. A noticeable increase in strength of these AlN materials after oxidation was not revealed.

A reduction in strength at room temperature from 500–600 to 250 MPa after oxidation above 1250° C was also observed for the materials produced by sintering the Al$_2$O$_3$/AlN mixture at 1800° C in nitrogen [5.32]. A major component of these materials is AlON, and the mass loss proceeds according to the reaction $4AlON + O_2 = 2Al_2O_3 + 2N_2$ to form α-Al$_2$O$_3$. Lower strength

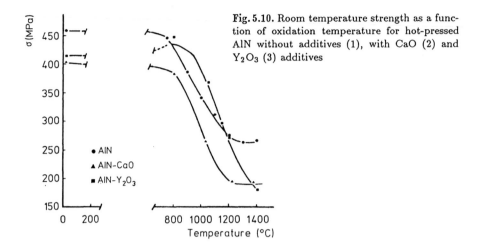

Fig. 5.10. Room temperature strength as a function of oxidation temperature for hot-pressed AlN without additives (1), with CaO (2) and Y_2O_3 (3) additives

after oxidation above 1250° C is due to pores formed along the grain boundaries in the surface layer of the material. In contrast to this, the exposure at 1050° C for 72 h results in a 40% increase in strength.

5.1.4 Zirconia Ceramics

Exposure in air at high temperatures leads to a change of properties of zirconia-based ceramics (Table 5.4), as in the case of nonoxide ceramics. Though the observed strength degradation is more extensive than for Si_3N_4 and SiC ceramics, this can be considered only as relative oxidation.

Aging at high temperatures leads to the tetragonal-to-monoclinic transformation in ZrO_2 and to a corresponding drop of strength, failure strain and thermal shock resistance of ceramics. Lower strength, even after short-time heating in an oxidizing atmosphere, was observed for ceramics manufactured by HIP in an inert medium. Such ceramics in their initial state are usually of dark colour because of defects in the oxygen sublattice of ZrO_2 and of lower stability than ZrO_2-based materials sintered in oxygen or in air. In this case the oxidation of ZrO_{2-x} to ZrO_2 does really take place.

5.2 Effect of Corrosion in Different Environments

5.2.1 Hot Corrosion

The information on the effect of oxidation in air on the properties of ceramics used for manufacturing different components of gas turbine engines and some other devices is insufficient, since during operation these components are also affected by various salts. As is shown in Sect. 3.1, salt-assisted hot corrosion changes the composition and structure of the surface layer and increases the concentration of defects. Therefore, the effect of salts on the mechanical proper-

Table 5.4. Oxidation behaviour of commercial zirconia ceramics [5.2]

Material	Properties as-received		
	Bending strength, MPa	Failure strain, %	Elastic modulus, GPa
Nilsen MS PSZ (MgO)	637	0.39	187
Feldmühle ZN-40 ZrO$_2$ (MgO)	426	0.23	193
Coors TT-ZrO$_2$ (3%MgO)	481	0.25	193
NGK Z-191 (Y$_2$O$_3$)	889	0.46	189
Nilsen TS PSZ (MgO)	523		
	Properties after 1000° C/1000 h exposure		
	Bending strength,* MPa	Failure strain, %	Elastic modulus, GPa
Nilsen MS PSZ (MgO)	271(−57%)	0.15	186
Feldmühle ZN-40 ZrO$_2$ (MgO)	298(−30%)	0.16	187
Coors TT-ZrO$_2$ (3%MgO)	146(−70%)	0.09	160
NGK Z-191 (Y$_2$O$_3$)	826(−7%)	0.43	189
Nilsen TS PSZ (MgO)	330(−36%)		
	Critial quench temperature difference, ΔT_c, ° C		
	As-received	After 1000° C/1000 h exposure	
Nilsen MS PSZ (MgO)	375	275	
Feldmühle ZN-40 ZrO$_2$ (MgO)	325	275	
Coors TT-ZrO$_2$ (3%MgO)	350	†	
NGK Z-191 (Y$_2$O$_3$)	400	275–300	
Nilsen TS PSZ (MgO)	425–450	275	

* Numbers in brackets indicate percentage change. † Coors material judged too much changed after 1000° C/1000 h exposure to definitively measure ΔT_c

ties of ceramics must be elucidated. The literature provides rather contradictory data on this problem. Thus, RBSN specimens whose surface is covered with sea salt are reported to corrode inconsiderably in air at 750° and 830° C, which slightly influences their strength [5.33]. As is shown in [5.34], cyclic and static tests of specimens in the stream of combustion products with the addition of sea salt at 900°–1150° C involve considerable strength degradation of RBSN and HPSN.

The tests of gas turbine blades made of NKKKM–83 ceramics in the stream of kerosene combustion products (Sect. 3.1.1) have demonstrated that at 1250°–1375° C many blades fail affected by thermal stresses and oxidizing atmospheres (Table 5.5).

Before the tests the lot of blades (25 pcs) was divided into five groups (5 pcs each). The first group included blades which revealed no defects in X-ray non-destructive flaw detection. The rest of them was oxidized or quenched at different temperatures before the tests (Table 5.5). The inspection of blades after the tests demonstrated that the blades of groups 1 and 3 turned out to be the least damaged ones. Oxidation appeared to be quite ineffective for heal-

Table 5.5. Visual inspection of blades after burner rig tests

Number of group	Pretreatment of blades	Condition of blades	State of specimen surface
1	Without treatment	Flawless (4) Spall (1)	Light grey White beads, scattered pits up to 0.2 mm in diameter
2	Oxidation at 900° C for 3 h	Flawless (2) Cracks (1) Spalls (2)	Light grey Many pits up to 0.2 mm in diameter
3	Quenching at $\Delta T = 300°$ C	Flawless (4) Spalls (1)	Dark grey Pits 0.5 mm in diameter, white beads on several blades
4	Quenching at $\Delta T = 800°$ C	Flawless (2) Spalls (3)	Dark grey White beads
5	Quenching at $\Delta T = 800°$ C followed by oxidation at 1350° C for 3 h	Flawless (1) Crack (1) Fracture (1) Spalls (2)	Light grey Scattered pits up to 0.2 mm in diameter White and yellow beads, network of small cracks on the surface

Note. The number in brackets is the number of tested blades

ing cracks arising during quenching of blades with a temperature difference exceeding ΔT_c.

We tested the strength and fracture toughness of bars cut from the turbine blades made of NKKKM–84 ceramics to determine their serviceability after thermocyclic loading during tests in a burner rig in the stream of diesel fuel (Table 3.2) combustion products at lower temperatures which did not result in a noticeable damage of the blade surfaces (Sect. 3.1.1). We also prepared bars from as-received blades to get comparative data.

Bars $3 \times 4 \times 25$ mm in size were cut from the surface (group A) and inner (group B) layers of blades. In the first case one surface of the bars was not machined, the other surfaces were ground. In the second case all the surfaces of the bars were ground. During bending tests the bars of group A were set with the as-received surface positioned in the tensile region. After the burner rig tests the strength of ceramics does virtually not degrade (Table 5.6). Moreover, the healing of surface flaws due to oxidation resulted in higher strength of specimens cut from the surface layer. Thus, in certain temperature ranges, silicon nitride ceramics can successfully operate even in combustion products of low-grade fuels.

The addition of sodium to fuels promotes corrosion and strength degradation of Si_3N_4 and SiC ceramics (Table 5.7). Extensive pitting under these conditions causes strength reductions of $\sim 30\%$ in three types of SiC and $\sim 50\%$ in siliconized SiC. Chemical attack of grain boundaries and the formation of pits result in a strength reduction of $\sim 30\%$ for the three Si_3N_4 materials studied. Because deposition is a continuous linear process, it is logical to assume

Table 5.6. Test results for bars cut from turbine blades

Bar type	Group	σ, MPa	K_{1c}, MPa $\cdot$ m$^{1/2}$
From as-received blades			
Blade 1	A	123/153	–
	B	225/175	3.0
Blade 2	A	156/213	–
	B	210/208	3.2
From tested blades			
1000 cycles	A	181/200	–
	B	157/168	3.0
500 cycles	A	173/208	–
	B	–/194	2.8

Note. The values at 20° C and 1200° C are given in numerator and denominator, respectively

Table 5.7. Room-temperature strengths of as-received and corroded SiC and Si$_3$N$_4$ ceramics [5.35]

Material source	Major additives	MOR		Strength reduction, %
		As-received	After corrosion	
Silicon carbide				
Carborundum Hexoloy SASC	B, C	401 ± 55	308 ± 42	23
General Electric β-SiC	B, C	406 ± 27	297 ± 50	27
Kyocera SC 201	B	451 ± 31	299 ± 46	34
Carborundum KX01	Si	431 ± 61	206 ± 94	52
Silicon nitride				
GTE AY6	Al, Y	668 ± 36	509 ± 78	24
Toshiba SSN	Al, Y	872 ± 67	627 ± 103	28
NGK SN50 SSN	Al, Mg, W, Zr	814 ± 55	547 ± 116	33

that very long exposure to these conditions would result in total consumption of the SiC and Si$_3$N$_4$ materials.

At the same time for other type of SiSiC [5.36], sulfur compounds (SO_2, H_2S) present in the operating environment do not affect the strength.

To simulate the effect of hot corrosion on strength and fracture toughness, the specimens of NKKKM–81 [5.37] and NKKKM–84 [5.38] ceramics (see also Sect. 2.1.4) were impregnated with a concentrated solution of a corresponding salt and dried at 120° C. Then they were mounted in the furnace of a testing machine and heated up to 1200° C. The specimens were exposed to this temperature for 2.5 h, then a part of them was tested at this temperature, and the rest was slowly cooled together with the furnace and tested at room temperature.

As opposed to oxidation of ceramics in air, salt-assisted oxidation reduces strength both at higher temperatures and at 20° C (Table 5.8). The drop of strength determined at 20° C becomes more and more pronounced with increasing oxidation temperature (Fig. 5.11).

Salts have a similar influence on the strength of both NKKKM–81 and NKKKM–84 ceramics (Table 5.8), which is characteristic of the materials of this type, despite the differences in the purity of raw materials and in processing.

Table 5.8. Oxidation of ceramics at 1200° C

| Treatment | NKKKM–81 | | | NKKKM–84 |
	σ, MPa	K_{1c}, MPa · m$^{1/2}$	a, μm	σ, MPa
Oxidation				
Salt-free	207/177	3.21/3.19	204	199/182
With				
NaCl	163/131	–/2.71	269	178/177
Sea salt	167/110	–/2.57	342	157/158
Na$_2$SO$_4$	177/93	–/2.38	411	195/134
Na$_2$SO$_4$ + V$_2$O$_5$	–	–	–	172/126

Note. The values at 20° C and 1200° C are given in numerator and denominator, respectively

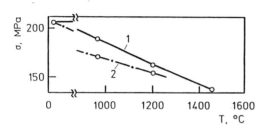

Fig. 5.11. Strength of NKKKM–81 specimens at 20° C as a function of salt-assisted oxidation temperature with NaCl (1) and sea salt (2)

However, these differences have an effect on the extent of salt-affected strength degradation. Thus, the strength of NKKKM–84 ceramics at 1200° C degraded to a lesser extent than that of the NKKKM–81 ceramics.

Since fuel used for marine-based gas turbines contains vanadium (up to 0.1%), apart from sodium and sulfur, an equimolar Na$_2$SO$_4$/V$_2$O$_5$ mixture was used to study the effect of oxidation on the mechanical characteristics of NKKKM–84 ceramics. This mixture degrades the strength even to a greater extent than Na$_2$SO$_4$ (Table 5.8). Lower strength of ceramics at room temperature after salt-assisted oxidation is caused by the formation of a rough, knobby oxide layer with multiple pits, cracks and other flaws (Figs. 3.3,4). The mechanism of strength degradation caused by a higher concentration of defects in the surface layer also explains the closeness of strength values at 20° C for both modifications after their salt-assisted oxidation, since defects formed in the oxide layer but not inherent in the initial structure of ceramics become a strength-controlling factor.

The calculated critical flaw size a determined by the equation

$$a = \frac{\pi}{5} \left(\frac{K_{1c}}{\sigma} \right)^2 , \qquad (5.1)$$

where σ is the strength, K_{1c} is the critical stress intensity factor, is in good agreement with the value measured experimentally (Table 5.8). This shows that other processes (besides formation of surface flaws) do not have any impact on further strength degradation of ceramics after oxidation.

As is seen in Fig. 5.12, salt-assisted oxidation also leads to a strength degradation of denser sintered ceramics, with the highest degradation occurring at

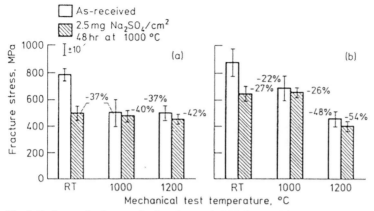

Fig. 5.12. Strength of as-received and corroded SSN materials at room temperature, 1000° C and 1200° C: SSN ($Al_2O_3 \cdot Y_2O_3$) (a) and SSN (Y_2O_3) (b) [5.39]

room temperature when the state of the specimen surface is of more importance.

Silicon carbide and silicon nitride ceramics display similar strength degradation during sodium salt-assisted oxidation (Table 5.9), the extent of degradation being also dependent on the depth and uniformity of pit distribution in the surface layer of specimens (see also Sect. 3.1.2).

The relation between the strength and the depth of pits analysed by fracture mechanics methods shows that K_{1c} for corroded specimens should be 2.6 MPa · $m^{1/2}$. The calculated K_{1c} nearly coincided with the value determined experimentally.

The strength of sintered α-SiC specimens coated with sea salt and oxidized in air below 1620° C was also investigated in [5.41]. Sea-salt–assisted oxidation at 960° and 1200° C was established to increase the strength of as-sintered specimens at room temperature due to healing of surface flaws. However, salt-free oxidation leads to a slightly greater increase in strength. Above 1200° C, just as in the case of Si_3N_4 ceramics, a salt present on the specimen surface involves active subcritical crack growth and a considerable decrease of strength at low loading rates.

As is seen in Fig. 5.13, oxide ceramics display a higher hot corrosion resistance than nonoxide ones. It was found that commercially available Ce–TZP

Table 5.9. Effect of hot corrosion on room-temperature bending strength and pit size distribution of SiC [5.40]

Salt corrodent	Strength, MPa	Number of pits per 1 mm² area		
		$D < 10\,\mu m$	$10\,\mu m < D < 20\,\mu m$	$D > 20\,\mu m$
As-received	409 ± 62	–	–	–
Na_2SO_4/SO_3	207 ± 72	1900	620	90
Na_2CO_3/CO_3	355 ± 70	2300	30	4
Na_2SO_4/air	251 ± 45	17300	120	6

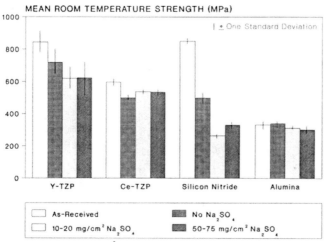

MEAN ROOM TEMPERATURE STRENGTH (MPa)

As-Received
10-20 mg/cm² Na₂SO₄
No Na₂SO₄
50-75 mg/cm² Na₂SO₄

Treated for 500 hours at 1000°C

Fig. 5.13. Mean strength after exposure with and without sodium sulfate. Treated for 500 h at 1000° C [5.42]

(Ceramatec) has excellent resistance to strength degradation by molten sodium sulfate compared to Y–TZP (NGK Insulators) or Si_3N_4. A possible mechanism of strength degradation in Y–TZP is the Y_2O_3 depletion of the Y–TZP surface. This would allow the tetragonal-to-monoclinic transformation of the zirconia to occur spontaneously, resulting in a reduction of strength that is usually attributed only to overageing.

The principal conclusions concerning the effect of 100 h Diesel engine combustion environment on PSZ ceramics (Nilcra Ceramics, USA) are as follows [5.43]:

- Exposure of Mg–PSZ of TS grade caused a substantial decrease (32%) in bending strength. This decrease was attributed primarily to an increase in monoclinic phase content as a result of exposure. Surface deposits and thermally induced or environmentally assisted microcrack growth may also have been important.
- Exposure of Mg–PSZ of MS grade caused only a 9% decrease in mean bending strength, which was also attributed to an increase in the content of monoclinic phase due to exposure.

High-purity, fully dense alumina is very resistant to hot corrosion (Fig. 5.13). However, it does exhibit a slight strength reduction which may be attributed to the corrosion of traces of a glassy grain-boundary phase.

There is quite a lot of information on the effects of coal slags on the strength of various ceramics [5.39, 44, 45]. The extent of strength degradation appears to be quite dependent on the type of ceramic and the slag, and it is difficult to identify any general trends. As mentioned, in an acidic slag only limited dissolution occurs, and either pitting or slag penetration may lead to strength degradation. Basic slags lead to rapid rates of material consumption and may also exhibit pitting. In general, whenever pitting was observed, strength degradations were also observed.

5.2.2 Molten Salt Corrosion

Corrosion in melts, as it was shown in Sect. 4.2, as well as hot corrosion can result in a considerable damage of the specimen surfaces. This leads to the degradation of the ceramics [5.46].

To elucidate the causes of strength degradation, HPSN (NC–132, Norton Co.) and RBSN (NC–350, Norton Co.) specimens exposed to molten NaCl and the 37% NaCl – 63% Na_2SO_4 eutectic at 1000° C were tested by four-point bending [5.34]. The specimens of NC–132 were exposed to melts for 144 h and those of NC–350 for 50 h. The results are summarized in Tables 5.10, 11. The glassy layer formed on the specimen surfaces after corrosion in molten NaCl was removed before the tests by a 15 min exposure in a 10% hydrofluoric acid solution.

Molten salt corrosion reduces the elastic properties of both materials and decreases the density of HPSN specimens due to etching of their structure with the salt mixture (Table 5.10). After molten salt corrosion the strength and fracture toughness of the materials were also deteriorated. Thus, at all the test temperatures the strength of HPSN reduced by $\sim 30\%$ after corrosion in NaCl and by $\sim 50\%$ after corrosion in the eutectic. The strength of RBSN was reduced approximately by one half after corrosion in both melts. The critical stress intensity factor determined on the Knoop-indented specimens by a procedure described in [5.45] decreased by $\sim 20\%$ for hot-pressed materials and remained almost unchanged for reaction-bonded ones. This confirms the hypothesis that lower strength is connected with a higher concentration of flaws in the corroded surface. Since in fracture toughness tests the fracture originated from a preformed defect, molten salt corrosion did not involve the reduction of K_{1c}.

Corrosion in melts gives rise to a marked increase (Table 5.11) in the critical flaw size determined by equation (5.1) from the test results, especially at 1000° and 1200° C. Thus, at the above temperatures the subcritical crack growth in corroded ceramics is likely to occur. To estimate if this process is possible at 1200° C, the specimens were tested at loading rates differing by two orders.

Table 5.10. Effect of molten salt corrosion on the physical properties of silicon nitride ceramics [5.34]

Treatment	E_d, GPa	G, GPa	μ	Density, kg/m^3
Untreated	313/181	121/74.4	0.28/0.21	$3200 \pm 12/2480 \pm 55$
Corrosion in melts				
NaCl	264/165	105/68.5	0.24/0.21	$3090 \pm 9/2530 \pm 11$
37% NaCl + 63% Na_2SO_4	182/180	74.1/75.8	0.23/0.19	$3050 \pm 8/2480 \pm 15$

Note. The values for NC–132 and NC–350 are given in numerator and denominator, respectively

Table 5.11. Effect of molten salt corrosion on the mechanical properties of silicon nitride ceramics [5.34]

	T, °C	Untreated	Corrosion in melts	
			NaCl	37% NaCl + 63% Na$_2$SO$_4$
σ, MPa	20	$\dfrac{800 \pm 15}{245 \pm 60}$	$\dfrac{540 \pm 32}{107 \pm 10}$	$\dfrac{305 \pm 100}{111 \pm 8}$
	800	$\dfrac{610 \pm 100}{143 \pm 28}$	$\dfrac{410 \pm 70}{79 \pm 8}$	$\dfrac{130 \pm 35}{74 \pm 7}$
	1000	$\dfrac{510 \pm 85}{175 \pm 40}$	$\dfrac{345 \pm 60}{90 \pm 23}$	150 ± 50 —
	1200	$\dfrac{520 \pm 55}{253 \pm 50}$	$\dfrac{375 \pm 60}{119 \pm 13}$	$\dfrac{235 \pm 30}{138 \pm 13}$
K_{1c}, MPa · m$^{1/2}$	20	$\dfrac{4.95 \pm 0.35}{1.63 \pm 0.07}$	$\dfrac{3.95 \pm 0.35}{1.76 \pm 0.35}$	$\dfrac{4.20 \pm 0.45}{1.63 \pm 0.09}$
	800	$\dfrac{3.80 \pm 0.25}{-}$	$\dfrac{3.15 \pm 0.25}{-}$	$\dfrac{2.15 \pm 0.12}{-}$
	1000	$\dfrac{3.70 \pm 0.62}{1.87 \pm 0.32}$	$\dfrac{3.59 \pm 0.40}{1.68 \pm 0.15}$	$\dfrac{3.20 \pm 0.12}{-}$
	1200	$\dfrac{5.93 \pm 1.4}{2.43 \pm 0.21}$	$\dfrac{4.55 \pm 0.25}{2.37 \pm 0.43}$	$\dfrac{4.65 \pm 0.32}{2.21 \pm 0.27}$
a, µm	20	$\dfrac{22}{28}$	$\dfrac{34}{169}$	$\dfrac{66}{135}$
	800	$\dfrac{27}{-}$	$\dfrac{38}{-}$	$\dfrac{87}{-}$
	1000	$\dfrac{32}{72}$	$\dfrac{70}{219}$	$\dfrac{108}{-}$
	1200	$\dfrac{80}{58}$	$\dfrac{98}{249}$	$\dfrac{222}{161}$

Note. The values for NC–132 and NC–350 are given in numerator and denominator, respectively

A considerable increase in strength with a loading rate was revealed [5.34]. The data confirm that subcritical crack growth does occur under the above conditions.

The investigation of fracture surfaces by AES has demonstrated that corrosion involves the diffusion of sodium along the pores and grain boundaries in the interior of the material. The fracture surfaces are heavily contaminated with salts and impurities present in the materials.

The authors of [5.48] investigated the effect of salt-assisted corrosion on the mechanical characteristics of NKKKM ceramics. The experiments to study the effect of corrosion on strength and fracture toughness were mainly performed with molten NaCl. As it could be expected, after the treatment of NKKKM ceramics with molten NaCl, their strength and fracture toughness at room temperature change inconsiderably. However, at 1200° C these characteristics are greatly impaired. The time of exposure to molten salts (Fig. 5.14) and the time of oxidation in air (Fig. 5.1) have almost no influence on the strength at both test temperatures.

It should be noted that the reduction in strength and fracture toughness after the salt-assisted oxidation of specimens tested at 1200° C (Table 5.8) was the same as in the case of the specimens treated in molten NaCl and tested at this temperature. Besides, one could follow the correspondence of strengths at

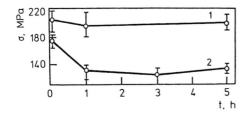

Fig. 5.14. Strength of NKKKM–81 ceramics at 20° (1) and 1200° C (2) as a function of time of exposure to molten NaCl

room temperature for the specimens corroded in molten NaCl at 950° C and oxidized at the same temperature in the presence of the above salt (Fig. 5.14). This points to similar mechanisms of the effect of different treatments on the mechanical properties of the ceramics under study in the same salt at the same temperature.

As opposed to Na_2SO_4, lithium sulfate mixed with LiF and Li [5.49] causes uniform corrosion of sintered α-SiC and lower strength degradation than LiS/Li/LiF and LiCl/Li/LiF mixtures. Above 600° C the latter mixtures brought about a reduction in strength by approximately one half.

5.2.3 Corrosion in Solutions

Corrosion in solutions, as corrosion in molten salts, leads to lower strength in those cases when the specimen surfaces are etched to form pits and slits on the grain boundaries. For example, the leaching of silicon from self-bonded SiC in an alkaline solution results in a 25%–30% reduction of strength [5.50]. On the other hand, uniform etching of the surface involving the dissolution of a defective surface layer of the material can even cause a certain increase in strength, as it was observed on etching of the SiC–TiB_2 ceramics in aqua regia [5.29].

6. Mechanical Properties and Corrosion

The influence of the environment is one of the reasons preventing a successful application of nonoxide ceramics in different high-temperature devices. Nevertheless mechanical tests on them are often performed only in inert media. According to the present knowledge [6.1, 2] the strength of ceramics on heating should slightly change first within the temperature range of brittle fracture, then increase somewhat when the temperature of brittle–ductile transition ($\sim 0.5 - 0.6\,T_m$) is approached, and at last decrease sharply in the temperature range of viscous flow (Fig. 6.1).

During the test in air or in other environments we can observe certain deviations from this relationship due to material-environment interactions. Though investigations in inert media give information on the real physical behaviour of ceramics, it cannot be used for estimating their performance under service conditions. Therefore, Si_3N_4- and SiC-based materials are mostly tested in air. But for B_4C and other materials, which still find limited application at high temperatures, practically all the data were obtained in inert media [6.3]. To describe the high-temperature behaviour of materials comprehensively, they should be investigated in inert media and in actual environments. Thereafter we have to analyse the mechanical behaviour of ceramics in various environments.

6.1 Effect of Oxidation on Results of High-Temperature Mechanical Tests

6.1.1 Silicon Nitride Ceramics

Oxidation exerts an particularly pronounced effect on porous reaction-bonded materials which are actively oxidized at the initial stage of heating, i.e. during the time necessary for short-term mechanical tests at high temperatures. Thus, the temperature dependences of strength, elastic moduli and critical stress intensity factors obtained in [6.4] for RBSN in nitrogen and air are different (Fig. 6.2). The factors responsible for strength variations of similar materials after oxidation were discussed in Chap. 5. The growth of elastic moduli in air is connected with an increase in the specimen density after oxidation. The growth of K_{1c} and γ_{ef} is determined by the formation of an oxide phase characterized not by brittle but by ductile fracture in the temperature range under consideration as well as by possible blunting of a stress concentrator as a result of oxidation

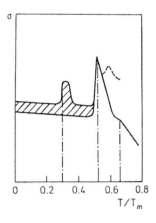

Fig. 6.1. Generalized plot of strength vs temperature for covalent polycrystalline materials in an inert medium [6.2]

and a SiO_2 layer formed at the notch tip. Thus, during the tests in nitrogen an increase in fracture toughness and a decrease of strength at 1400° C are caused by ductility appearing in the grain-boundary phase.

Comparative tests of NKKKM–84 ceramics in air and argon (Fig. 6.3a) also revealed the difference between σ values, but only in the range of intermediate temperatures. A similar effect of oxidation was also observed in the tests of Y_2O_3-doped HPSN [6.5] and $Y_2O_3 + Al_2O_3$-doped HPSN in air (Fig. 6.3b). The strength degradation of these ceramics, in which the grain-boundary phase in the Y–Si–O–N system is formed, was caused by the cracking of the specimen surfaces (Fig. 5.6) due to oxidation of silicon-yttrium oxynitrides during heating up to test temperature [6.6]. Similar TiN-doped ceramics with a somewhat

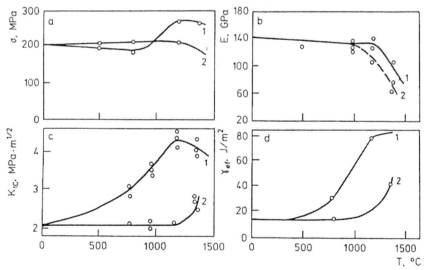

Fig. 6.2. Four-point bending strength (a), elastic moduli (b), critical stress intensity factors (c) and effective surface energy (d) as a function of temperature for RBSN in air (1) and nitrogen (2) [6.4]

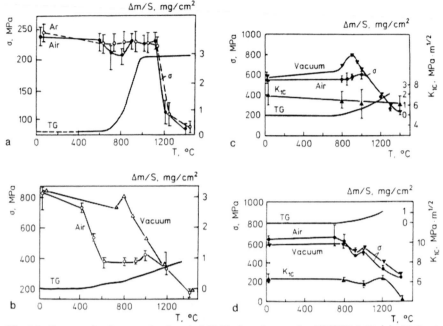

Fig. 6.3. Test results for reaction-bonded Si_3N_4-based ceramics NKKKM–84 (a), $Y_2O_3 +$ Al_2O_3-doped HPSN (b), HPSN of the same composition with 10% TiN (c) and MgO-doped HPSN (d)

modified composition of the grain-boundary phase did not display a strength loss during tests in air (Fig. 6.3c), but the strength in an inert medium over the range of intermediate temperatures was nevertheless higher. It should also be noted that for MgO-containing HPSN, with the grain-boundary phase exhibiting ductility at the lowest temperatures, the difference between temperature dependences of strength in vacuum and in air was quite inconsiderable (Fig. 6.3d).

Possible reasons for strength degradation of silicon nitride ceramics in the range of intermediate temperatures will be discussed in Sect. 6.3. Here we should note that at higher temperatures, when the grain-boundary phase in doped silicon nitride ceramics starts softening, the strength of specimens is almost the same in inert and oxidizing media. It is not the presence of flaws in the surface layer of specimens but viscous flow of the grain-boundary phase that becomes a strength-controlling factor.

Oxidation has even greater influence on the test results which require long-term exposure of specimens at high temperatures. Thus, oxidation abruptly impairs creep resistance [6.7]. The deformation of silicon nitride ceramics during creep tests is controlled by grain-boundary sliding and, thus, is determined by the state of the secondary grain-boundary phase. Therefore, β-sialons, SiMeON and similar materials which have almost no secondary phase along the grain boundaries display the highest creep resistance. Fine-grained hot-pressed materials containing oxide additives (MgO, CaO, MgAl$_2$O$_4$ and others) are less resistant. And an increase in MgO contents in Si$_3$N$_4$ results in the growth of strain values. Thus, for dense materials there is a certain correlation between oxidation resistance and creep. The materials with a good oxidation resistance (Chap. 2) are also resistant to creep; higher contents of CaO, SrO, MgO increase creep rates and mass gain on oxidation.

The studies on creep of dense silicon nitride ceramics produced by different methods [6.8] revealed a parallelism in the temperature dependence of creep and oxidation, and a characteristic temperature T_t for each individual material, which is decisive for its performance, was determined. At this transition temperature T_t the first occurrence of liquid phases in the microstructure is

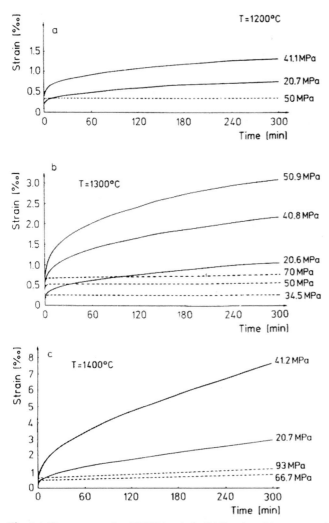

Fig. 6.4. Creep curves for RBSN in air (solid lines) and in vacuum (dashed lines)

indicated, which has serious effects on the properties. These temperatures were measured as 1460° C for HPSN (Y), 1240° C for SSN (Y, Al and Ti) and 1300° C for SRBSN (Y, Al and Mg). The oxidation kinetics is changing considerably at these temperatures. The nature of the creep curve is transient below these temperatures, while beyond the transition temperatures rather short rupture times are measured with preceding creep stages, with minimum creep rates and accelerated creep.

The comparison of creep in air and in an inert medium at 1400° C shows that the creep rate in air is lower. Oxidation is known (Sect. 2.1) to cause a gradual removal of the trace elements from the matrix into the oxide layer. This should give rise to a reduction in creep rates. In a reducing environment however, the trace elements will remain within the structure and higher creep rates will persist throughout the experiment, as has indeed been observed.

The situation with reaction-bonded ceramics is somewhat different. In reducing and inert media or in vacuum RBSN displays a higher creep resistance than most hot-pressed materials. However, the heating of materials with a density of $2110-2260\,kg/m^3$ in air (Fig. 6.4) or in oxygen leads to their inner oxidation, which gives rise to a viscous silicate phase along the grain boundaries that increases the creep rates by 1–2 orders [6.10]. For denser materials ($>2600\,kg/m^3$) the creep rate in air increases only by 20% compared with the results obtained in argon tests [6.11]. Hot-pressed materials with MgO addition have a lower creep resistance than the reaction-bonded ones [6.11] and are less sensitive to a test medium [6.10], since they are oxidized only on the surface.

6.1.2 Silicon Carbide Ceramics

Silicon carbide ceramics find wide application in high-temperature devices. Therefore, their high-temperature strength as well as the effect of oxidation on their properties have been studied thoroughly enough [6.12–14]. The researchers investigated strength, subcritical crack growth and/or deformability, which often contributed to understanding fracture mechanisms.

Silicon carbide should retain an almost constant strength up to high temperatures (its temperature of the brittle-ductile transition is 2000° C [6.15]), which corresponds to the results obtained for recrystallized SiC [6.16] and hot-pressed SiC with B_4C addition [6.12]. However, the strength of self-bonded SiC degrades at high temperatures (Fig. 6.5).

As is seen in Fig. 6.5, starting from 700° C the strength of specimens tested in an inert medium is much higher than that of specimens tested in air. In argon the strength decreases gradually up to 600° C, at 700° C it increases abruptly due to the ductile–brittle transition in the silicon phase, then it goes gradually down again up to 1300° C and drops considerably at 1300°–1400° C due to silicon melting. This relation is in good agreement with the generalized temperature dependence shown in Fig. 6.1.

Already in the range of 800°–1000° C the surface of specimens tested in air revealed oxidation of inclusions of iron compounds and other impurities. This

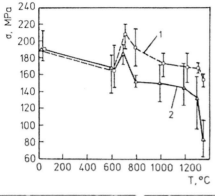

Fig. 6.5. Strength variation as a function of temperature in argon (1) and in air (2)

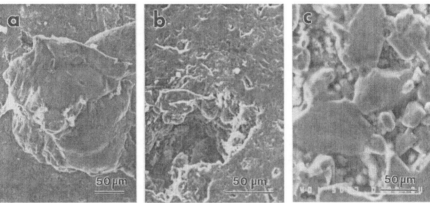

Fig. 6.6. Electron micrographs of specimen surfaces tested in air before (a) and after etching in hydrofluoric acid for 0.5 h (b,c): a,b – 1000° C, c – 1400° C

is accompanied by the blistering of oxidation products (Fig. 6.6a) and flaws about 200 μm in size appearing in these places, particularly well pronounced after mechanical removal or hydrofluoric acid dissolution of oxide blisters (Fig. 6.6b). Fractographic investigations demonstrated that precisely these flaws were the main fracture origins for the specimens tested at 800° and 1000° C.

At higher temperatures the major components of the ceramics, SiC and Si, start oxidizing. In this case iron, calcium, aluminium and other impurities, as it was shown earlier, diffuse to the surface and form a relatively low-melting silicate layer which is in the liquid state above 1200° C. Oxidation is the most active on impurity-enriched areas, along the grain boundaries etc. When the oxide layer is removed from the surface, one can see that oxidation results in grooves along the grain boundaries (Fig. 6.6c) and other flaws in the surface layer; they become stress concentrators and lead to a strength degradation of the material. The mechanism of crack initiation originating from the flaws formed in the surface layer during oxidation is schematically shown in Fig. 6.7.

The investigation of specimen fractures helped to establish the exact fracture mechanisms of the material in various environments. The fractures of

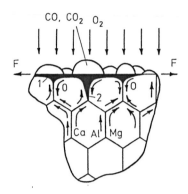

Fig. 6.7. Scheme of crack initiation on the specimen surface in the presence of a liquid silicate phase: liquid oxide phase (1); crack (2)

specimens tested in air up to 1300° C do not differ much from those obtained in an inert medium, except for the fact that they show oxidation traces in the range of 1100°–1300° C. At the same time the fractures of specimens tested at 1350° C and above appeared to be quite different in their nature. Due to oxidation or melting of silicon, these specimens adhered to the support, their fragments did not scatter after the propagation of a crack through the whole specimen and remained on the support, as it is schematically shown in Fig. 6.8.

During slow cooling of specimens together with the furnace, a freshly formed fracture surface was oxidized. Because of a very small crack opening displacement, the oxygen transport to the reaction surface and the removal of oxidation products (CO, CO_2) became the limiting stages of the oxidation process. As a result, a continuous oxide layer is formed only on the edges, and on a larger part of a fracture tension zone SiO_2 fibres are formed. The formation of silica whiskers was detected earlier during oxidation of self-bonded SiC in the gas stream at a low partial oxygen pressure [6.17].

On the edges of a specimen, in the area of larger crack opening, the fibres and whiskers were growing from a SiO_2 layer (Fig. 6.8c) and in the center directly from a smooth SiC surface (Fig. 6.8d,e). In the compression zone, where the crack opening was minimum, the partial oxygen pressure appeared to be insufficient for the growth of SiO_2 whiskers, and in this area gaseous SiO was probably formed. But as can be seen in Fig. 6.8e,f, the reactions leading to the SiO and SiO_2 transport through the gas phase and more active diffusion processes on the surface completely change the character of the fracture surface (cf. Fig. 6.8a).

The discovered effect suggests that the healing of cracks in the specimens and partial recovery of their strength [6.18, 19] can be connected not only with the filling of these cracks with melted silicon [6.18] or silica [6.19], but also with the mass transfer of silicon carbide through the gas phase the driving force of which is the reduction in the surface energy. Strength recovery in

Fig. 6.8. Scheme of fracture and fracture surfaces (a – f) of specimens tested at 1350° C in argon (a) and in air (b – f), in points 1(b,c), 2(d), 3(e), and 4(f)

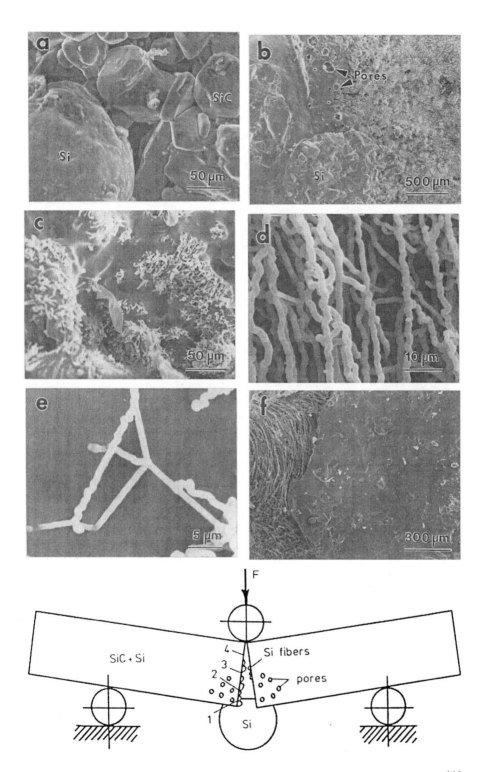

vacuum [6.19] does not occur, since the mass transfer through the gas phase is impossible without oxygen. When the crack does not cross the whole specimen, it is gradually blunted and closed.

Thus, oxidation can not only give rise to new flaws populations in the material, but can also promote the healing of surface cracks.

However, it should be noted that the strength degradation of the described ceramics at high temperatures both in an inert medium and in air was more extensive than, e.g., for an NC–435 material (Norton Co., USA) [6.18]. First of all, this is due to iron, calcium, aluminium and other impurities which reduce the oxidation resistance of ceramics. The removal of impurity inclusions would prevent a strength degradation due to pitting corrosion at 800°–1000° C.

The strength of recrystallized SiC does not decrease with temperature; moreover, it is likely to increase, e.g. for Crystar ceramics [6.16] because of their higher oxidation resistance and high-purity as well as because of the absence of free silicon and an easily softening phase along the grain boundaries. The oxidation of recrystallized ceramics generates an amorphous SiO_2 layer which heals flaws on the surface and defects in the inner layers of specimens, thus actually increasing their strength. These data confirm once more that ceramics cannot be considered simply as a compacted material, since their high-temperature behaviour is dependent on their processing.

The SiC products are generally very creep resistant; similarly, the oxidation resistance is good, if the content of doping elements and impurities is not too high. Under these conditions no distinct influence of preoxidation treatment on creep was found [6.20]. β-SiC and CVD-SiC are slightly less creep resistant than α-SiC, but reveal the possibility of relatively high creep deformation before fracture. This effect has not yet been fully explained; in the case of CVD-SiC it seems to be a consequence of the unique microstructure.

6.1.3 Boron Carbide Ceramics

The use of boron-carbide-based materials at high temperatures is limited at present, which probably results from the fact that they have been investigated primarily at room temperature [6.3, 21].

In this section we present the results on mechanical behaviour of two types of hot-pressed boron carbides [6.22–24] which differ insignificantly in their characteristics from those considered in Sect. 2.4.

As can be seen from Fig. 6.9a, the temperature dependences of strength in air and in argon differ significantly. In argon at a temperature above 1000° C the strength of the specimens increases, while in tests in air it decreases. Fractographic investigations establised that the fracture of the material was brittle in the whole temperature range studied. Fractographic analysis of the specimens tested at elevated temperatures in air was complicated by the formation of an oxide layer. In order to remove it, the specimens were boiled for 10 min in distilled water. Despite the fact that oxidation leads to the etching of the fracture surfaces, it was possible to establish that the fracture origins were primarily craters formed in the surface layer (Fig. 2.44b,c). Investigations after

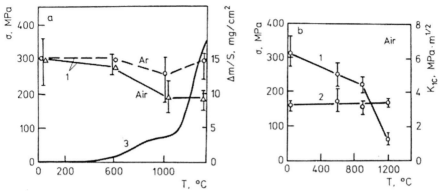

Fig. 6.9. Strength (1), fracture toughness (2) as a function of temperature and TG curve (3) for hot-pressed B_4C with Al and Si (a) and without additives (b)

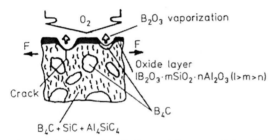

Fig. 6.10. Scheme of oxidation and fracture of the material for the tests in air above 700° C

removal of the oxide layer from the surfaces of specimens tested at 1000° and 1400° C in air make it possible to confirm that the formation of craters occurs specifically in the areas of pure boron carbide (Fig. 6.10).

During the tests in argon no oxide layer was formed on the surface of the specimens. XRD detected only very weak peaks, which can be attributed to free boron appearing on the dissociation of B_4C or to lower oxides of boron. However, the formation of a relief visible to the naked eye on the surface of specimens tested at 1400° C indicates that the argon used by the authors, as in [6.25], was not sufficiently pure. Oxidation of B_4C, which occurred at a low partial oxygen pressure, led to the formation of gaseous boron oxides and removal of them from the surface of the specimen. In this case significant oxidation was noted only at 1400° C. The depth of etching of the specimens was less than $10\,\mu$m, that is, many times less than on oxidation in air.

The presented results of investigations of specimens after tests and also the data on the oxidation process of the material on heating in air (Sect. 2.4) explain the character of the changes in strengths with temperature given in Fig. 6.9. The relationship obtained in an inert atmosphere is typical of similar materials and is characterized by a gradual reduction in strength up to 1000° C and some increase at 1400° C which is apparently related to the brittle–ductile transition occurring at temperatures of $\sim 0.6\,T_m$ (Fig. 6.1) (the same charac-

121

ter of temperature relationship of the ultimate strength for hot-pressed boron carbide was observed by *de With* [6.25]). The reduction in the elastic moduli of the ceramics at 1200° C and above indicate the probability of this assumption, but it was not possible to record nonlinear deformation of the ceramics all the way up to 1400° C. In air at 600° C the values of the ultimate strength differ weakly from those obtained in argon, since pronounced oxidation of this material starts at a temperature above 700° C (Fig. 6.9a). At the same time with a further temperature increase the difference between the strength values obtained in argon and in air increases. This is related to activation of the oxidation process and an increase in the depth of the craters formed in the surface layer of the specimens (Fig. 6.10).

However, for the investigated boron carbide the reduction in strength was not as significant as for the material described in Fig. 6.9b. This may be due to the fact that in the second case on oxidation of material with a more uniform structure [6.22], etching of the grain boundaries occurs and deep grooves are formed, which on loading easily develop into cracks (Fig. 2.41). In the present case, oxidation-induced flaws have a larger radius of curvature and the reduction in strength is less, since the Ioffe effect can arise.

Fracture toughness tests did not reveal any noticeable decrease of K_{1c} in oxidizing media, since the fracture was induced by a previously made notch. Therefore a higher concentration of defects in the surface layer of the specimens did not influence the obtained results (Fig. 6.9b).

Thus, the silicon- and aluminium-doped material retains a higher strength in air at elevated temperatures than pure B_4C, due to its higher oxidation resistance.

It follows from the above data, that environmental effects (Fig. 6.11) can also cause the growth of strength, due to an increase in density and the healing of flaws on the specimen surface (which was observed for recrystallized and sintered SiC and RBSN), and due to the removal of impurities from ceramics and compression stresses arising in the surface layer of the material (which was revealed for HPSN).

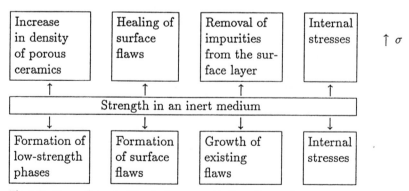

Fig. 6.11. Strength-controlling factors on oxidation of ceramics

However, much more often oxidation deteriorates the properties of ceramics. The reduction in strength may be connected with the phase transformation in the surface layer of specimens leading to the formation of low strength phases or to high internal stresses which can give rise to macroscopic strain of specimens or even their cracking and failure. Of course, such visually detectable defects are far from regular. Oxidation usually results in pitting, etching of grain boundaries, as it was observed for HPSN, self-bonded SiC, hot-pressed B_4C and others, as well as in the cracking of specimen surfaces. Here, together with new flaw populations usually appearing on the grain boundaries and inclusions of other phases, a growth of existing defects (pores and cracks) is possible. In these cases the bending strength of materials decreases both at high temperatures and at room temperature after heating in air up to an initial temperature of oxidation.

6.2 Effect of Salts on High-Temperature Strength

As it was already mentioned in Sect. 3.1, the service environments projected for structural ceramics are quite hostile. In a gas turbine engine some of the structural ceramics will face hot corrosion, i.e. attack by molten Na_2SO_4 which condenses on engine parts when injected NaCl reacts with sulfur impurities in the fuel. Marine propulsion gas turbine engines experience a similar sulfate-induced corrosion. Therefore, the serviceability of ceramics should be tested under conditions simulating the actual environments. The analysis of oxidation effect on high-temperature strength of ceramics (Sect. 6.1) and the information on surface damage by hot corrosion (Sect. 3.1) suggest that the strength of the material measured under conditions simulating service environments for engine ceramics (high temperature + sodium salts) will be lower than that measured in a neutral medium. However, it would be useful to compare the extent of environmental effects on the measured strength values using several materials as an example.

The comparison of the strength of reaction-bonded NKKKM–84 specimens [6.26] impregnated with salts and in the intial state has demonstrated that the strength of a salt-impregnated material degrades considerably and that this becomes most pronounced at 700°–1100° C, especially for the $NaCl/Na_2SO_4$ eutectic (Fig. 6.12). It is essential that the strength of NaCl- and eutectic-impregnated specimens deteriorates to a greater extent during tests in air than in a neutral medium (Fig. 6.12).

Analysing the obtained results, we have every reason to assume that above 1200° C the strength of the material under study in the presence of salts is mainly determined by the extent of softening of a glassy phase confined to the grain boundaries and influenced by the processing of these ceramics [6.26]. It was really established that lower strength of the material in this temperature range was accompanied by noticeable nonlinearity of strain and creep diagrams, but in the range of 1200°–1400° C the character of strength variations of initial

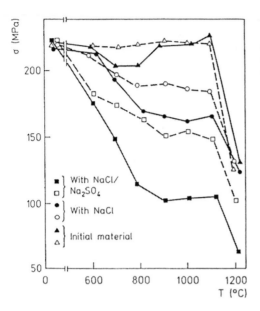

Fig. 6.12. Strength of $Na_2SO_4/NaCl$- and NaCl-impregnated NKKKM–84 specimens and initial NKKKM–84 specimens as a function of temperature during tests in air (solid lines) and in a neutral medium (dashed lines)

and salt-impregnated specimens is similar. Moreover, the NaCl impregnation did not influence the test results at all. A somewhat lower strength of eutectic-impregnated specimens can probably be attributed to pitting due to etching of the material (Fig. 3.4). The traces of etching were also well visible on the grain boundaries.

At 700°–1100° C one of the possible mechanisms of salt-induced strength degradation can be chemical corrosion of the material. This assumption is in agreement with the similarity of the plots of strength vs temperature for air and neutral media and with pitting arising at higher temperatures.

However, while the possibility of chemical corrosion for the eutectic (due to Na_2SO_4) can be explained theoretically and demonstrated in practice [6.27], for NaCl this cannot find any corroboration.

In this connection we must examine the effect of adsorption strength deterioration in the presence of a molten salt on the specimen surface [6.28]. However, the beginning of an abrupt drop of strength in the presence of NaCl and the eutectic corresponds to the same temperature and does not correlate with the melting temperature of these salts. Moreover, according to the TG data, most of the salt evaporates after melting. This is also confirmed by XRD data. Thus, we have every reason to believe that at 700°–1100° C the reduction in strength of the ceramics under study is determined by several factors. But even the whole complex of considered mechanisms does not fully explain the obtained results. First of all, this refers to a well-pronounced difference in experimental data for air and neutral media, which gives evidence of an important role of oxygen in the process of strength degradation. As it was reported in Sect. 3.1, the Na_2SO_4-induced corrosion of such ceramics is heavier in the presence of oxygen. We should also emphasize that an abrupt strength degradation occurs

Fig. 6.13. Typical SiO_2 whiskers on the surface of specimens tested in an inert medium at $1200°$ C

precisely at the temperatures corresponding to the start of oxidation. Therefore we may suppose that salts activate the Si_3N_4 oxidation due to the presence of sodium which, as is known [6.29], increases the diffusion coefficient of oxygen in oxidation products by several orders, which, in turn, decreases the resistance to further oxidation.

Thus, sodium salts facilitate oxidation of a larger volume of the material which makes grain bonding weaker and reduces strength. At the same time these salts abruptly decrease the melting temperature of oxidation products (consisting mainly of SiO_2), which results in a continuous oxide layer on the surface of the material. If this layer thick enough, it retards oxidation.

With an increase in test temperatures, at the expense of longer heating and higher oxidation rates, a well-defined and rather thick oxide layer formed on the surface stops the oxidation and strength degradation as a result.

In connection with the above considerations we should note that during the tests in an inert medium strength properties could be influenced by residual oxygen in argon and in pores of the material. The presence of such oxygen is confirmed by the formation of SiO_2 whiskers (Fig. 6.13) observed earlier in the investigations of silicon carbide ceramics at low partial pressures.

The plots of strength vs temperature for TiN-containing HPSN (Fig. 6.14) obtained in different environments are similar to the plots for NKKKM–84. Thus, we may speak of common mechanisms of salt-affected strength degradation for different silicon nitride materials.

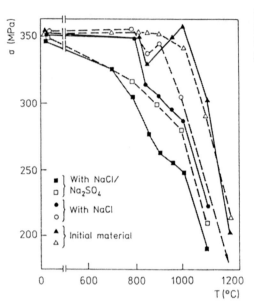

Fig. 6.14. Strength of Al_2O_3-doped HPSN–TiN as a function of temperature (designations correspond to those given in Fig. 6.12)

6.3 Failure of Ceramics Affected by Long-Term Mechanical Loading in Corrosive Environment

Generally speaking, the failure of materials can be divided into three large groups:

1. **Mechanical failure.** It occurs irrespective of the environment due to mechanical and thermal stresses.
2. **Corrosion failure.** It occurs irrespective of external stresses due to chemical interaction of the material with the environment.
3. **Corrosion-mechanical failure** (stress corrosion, stress corrosion cracking (SCC)). It occurs when the material is simultaneously affected by the corrosion environment and external loading under certain conditions.

In the previous sections only the two first cases of failure were described. Even on long-term loading in creep tests we examined only the strain of specimens but not their failure. The rest of the cases assumed that the processes of oxidation and mechanical failure (crack propagation in the material) were independent of each other. At the same time it is known that a simultaneous effect of corrosion environments and mechanical stresses results in SCC of metals and alloys [6.31]. For ceramics this effect has not yet been studied thoroughly enough.

The fact that the environment has an effect on the strength was observed for metals, minerals, glasses [6.28], oxide and nonoxide ceramics [6.32,33] and other materials. The selective character of this phenomenon relatively to different agents indicates that a chemical reaction of the material with the environment is a determining factor in all the cases.

The problem of stress corrosion resistance of ceramics has become particularly important in the manufacturing of components of engines and other high-temperature devices; for ensuring their serviceability, one should be able predict the behaviour of the materials during tens of thousands of hours from the data of short-term tests [6.34,35]. Though the methods for performance prediction of metals and glasses are quite elaborate, they cannot be applied to ceramics without any changes.

At the same time, ceramics have much in common with glass, which, owing to its homogeneity, is a good model to elucidate the mechanism of environmental effects on the strength of brittle materials at moderate temperatures. It should also be noted that silicon nitride-based ceramics, which are considered to be the most promising material for structural components, usually contain the secondary glassy phase of the system $MeO - SiO_2$ along the grain boundaries. Large amounts of glassy phase are also formed on oxidation of silicon nitride and silicon carbide materials during their operation.

For example, HPSN after oxidation, just as glass, becomes sensitive to loading rates in humid lab air (Table 6.1).

Therefore the theories of SCC developed for glasses (e.g., the Charles–Hillig theory [6.36]) can partially be applied to the above ceramics.

Recently attempts have also been made to develope improved theories to describe the failure of brittle materials (both glass and ceramics) affected by mechanical stresses and aggressive media [6.37–41]. The theories applicable to the conditions of H_2O vapour corrosion at temperatures close to room temperature [6.37] and the theories considering a diatomic gas as a chemically active medium [6.38, 39], which makes them applicable to the description of high-temperature SCC in various gaseous atmospheres, have been proposed by different scientists. *Thomson* and *Fuller* believe that the acceleration of subcritical crack growth is caused by an adsorption reduction in surface energy (Fig. 6.15). With the assumption of linear fracture mechanics that the crack

Table 6.1. Test results for HPSN with 5% Y_2O_5 and 2% Al_2O_3 in air at different loading rates

State of specimen	Temperature, ° C	Traverse speed, mm/min	MOR, MPa
Initial	20	5.5	695
Initial	20	0.008	705
After oxidation at 1250° C for			
3 h	20	5.5	715
3 h	20	0.008	632
Initial	800	5.5	682
Initial	800	0.008	713
Initial	1300	5.5	510
Initial	1300	0.008	164

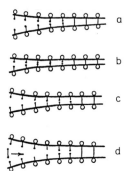

Fig. 6.15. Vacancy mechanism for atmospherically assisted fracture: The black circles represent adsorbed atoms and the dangling bonds represent vacancies. The crack grows when a bond breaks and the vacancies migrate out of the crack core, as shown in the sequence (b), (c). Finally in (d), the original configuration (a) is restored when new molecules chemisorb on the surface [6.38]

tip can be modelled as an elastic continuum, the authors of [6.39] derived the equation for the activation energy of subcritical crack growth which accounts for "chemical" and "mechanical" contributions to the activation energy value.

In contrast to this, in [6.40] the growth of stress corrosion cracks is presented as a cyclic two-stage process. At the first stage the bond at the tip of a sharp crack weakens, at the second stage the bond breaks due to simultaneous effects of mechanical stresses and random thermal fluctuations of atoms at the crack tip.

Gee and *McCartney* [6.41] proposed a statistical theory of failure of ceramics in aggressive environments which can serve as a basis for predicting the long-term behaviour of components by the results of conventional bending tests. However, according to this theory, one should know the distribution of flaws on the surface and in the interior of the material, the laws of subcritical crack growth rates etc. As the authors recognize, the absence of sufficient data on the mechanical behaviour and properties of the material makes it impossible to apply this theory in current practice.

The theories developed for the case of electro-chemical growth of corrosion-mechanical cracks in metals can also be applied to the description of SCC of ceramics in liquid media [6.42]. While most ceramic materials are dielectrics, neglecting electrical phenomena the process may be considered to proceed under diffusion conditions. The propagation of cracks in the specimens becomes possible only in those cases when a solid body possesses continuous-bonded regions, their solubility in liquids exceeding that of the other regions. This is the case, e.g. on oxidation of silicon nitride materials containing the secondary grain-boundary phase. A liquid oxide phase formed on the specimen surfaces above 1100° C interacts easily with the secondary silicate phase dissolving the latter. This results in grooves appearing along the grain boundaries of silicon nitride [6.43].

In all the discussed cases the crack propagation is a thermally activated process, since the crack growth rate is proportional to the diffusion coefficient or to the rate constant of the reaction (i.e. to thermally activated values). For the processes occurring at the crack tip, the relation of the crack growth rate to the stress intensity factor K_1 and absolute temperature takes the form

$V = f(K_1) \exp(-E/RT)$. The activation energy E corresponds to a specific physico-chemical process at the crack tip [6.42].

6.3.1 Silicon Nitride Ceramics

The data found in the literature on the sensitivity of Si_3N_4 ceramics to corrosion cracking are rather contradictory. The author of [6.44] believes that the slow crack growth in Si_3N_4 cannot be caused by corrosion phenomena, since the activation energy of the process is higher than the energy of a chemical reaction, all the attempts to promote crack growth at room temperature in the presence of different corrosive agents failed, investigations of crack resistance at high temperatures revealed the tendency to the closure of preformed cracks. At the same time many researchers present information on the sensitivity of silicon nitride ceramics to SCC.

In [6.33, 34] SCC of RBSN and HPSN was investigated at 20° and 300° C. The specimens were not ground and were tested by four-point bending at different loading rates and under static load in HCl and NH_4OH solutions, in distilled water, and in dry and humid air. The strength of specimens in humid air is 10% lower than in dry air. When pH decreases from 7 to 1, the strength of HPSN also decreases, but this effect is observed only at low loading rates.

Grinding which removes an impurity- and silica-enriched surface layer sharply decreases the sensitivity of the material to environmental effects. This caused the authors of [6.33] to conclude that the mechanism of stress corrosion at moderate temperatures is the ion exchange between the grain-boundary phase and H_2O. In the tests of specimens with a preformed crack (Knoop indentation) the value of strength was independent of the loading rate. However, if crack sizes are reduced and shifted somewhat from the centre of the specimen, in half of the cases failure originates not from a preformed crack but appears in a different area. Thus, one may suppose that corrosion of Si_3N_4 at temperatures close to room temperature exerts a greater influence on the formation of new flaws in the surface layer than on the growth of existing defects. The inapplicability of the theory of slow crack growth to the description of SCC of Si_3N_4 ceramics under the above conditions is also corroborated by the fact that the strength of specimens which did not fail during the tests for 1000 h was higher than the average strength of as-received specimens [6.33].

Oxidation of HPSN under a 160 MPa load at 1370° C in air for 1 h gives rise to a subcritical crack growth and to a reduction in strength as a result [6.45]. However, a slow crack growth at high temperatures can be caused not only by stress corrosion but also by physico-mechanical processes. Most literature data were obtained in static bending tests of Norton materials [6.46]. It was established that reaction-bonded materials exhibited a higher resistance to static fatigue than hot-pressed ceramics (Fig. 6.16). Their failure is retarded only under loads close to the value of ultimate strength.

During the tests above 1000° C ceramic degradation can occur due to nucleation of new flaws by surface reactions, crack growth and creep damage [6.47]. At moderate temperatures the latter mechanism of failure does not operate

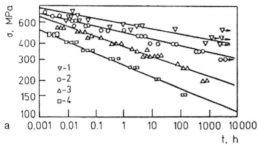

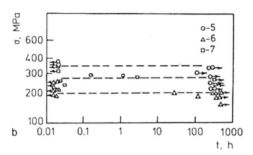

Fig. 6.16. Time to failure as a function of applied load for hot-pressed NC–132 (a) at T, °C: 1000(1), 1100(2), 1200(3), 1300(4) and RBSN (b): NC–350(5), KB1(6), Ford (density 2700 kg/m³) (7) at 1200° C [6.46]

and cannot explain the strength degradation of silicon nitiride materials (Fig. 6.3, 17) revealed [6.48–52] at intermediate temperatures. One may consider that this strength degradation is typical of silicon nitride materials, but the extent of this degradation and its temperature range are dependent on their composition and structure. Section 2.1.6 examined also the anomalies of oxidation kinetics of silicon nitride ceramics in this temperature range.

Even greater differences in strength are observed for the NKKKM–84 specimens tested at 800° C in air and argon under static load [6.53–55]. While in argon the specimens did not fail for 20 h (under the load of 86% MOR), in air the specimens could not even bear the load of 64% MOR (Fig. 6.18). At 800° C the specimens subjected to a static load for only 0.5 h exhibited a noticeable reduction of the residual strength, while at 20° C and 1100° C the strength of the specimens did not change (Table 6.2). It is noticeable that the oxidation of the unloaded specimens at 800° C also did not cause any changes in strength.

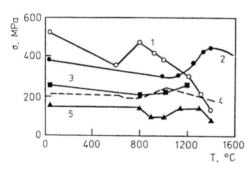

Fig. 6.17. Strength as a function of temperature for various silicon nitride materials tested in air: (1) [6.49]; (2) [6.50]; (3) [6.51]; (4) [6.48]; (5) [6.52]

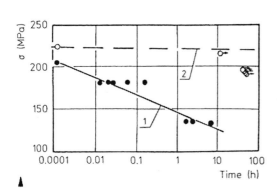
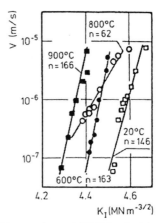

Fig. 6.18. Bending-stress–rupture results for the NKKKM–84 ceramics at 800° C in air (1) and in argon (2)

Fig. 6.19. Characteristic $K_1 - V$ diagrams of the NKKKM–84 at different temperatures

Table 6.2. Bending-stress–rupture results

Temperature, °C	MOR, MPa	Static load, MPa	Static load, MOR, %	Loading time, h	Residual strength, MPa
20	223	203	89	1	223
800	205	138	66	0.5	187
800	205	160	78	0.5	177
1100	224	203	89	0.5	221

The fracture behaviour is in good agreement with the activation of a subcritical crack growth in the temperature range of interest, which is accompanied by a considerable decrease of the slope of the $K_1 - V$ diagrams (Fig. 6.19) expressed by the exponent $n(V \sim K_1^n)$. Though we did not observe any noticeable changes in K_{1c}, the author of [6.56] did find a certain decrease of this parameter when they investigated HPSN with Y_2O_3 and $Y_2O_3 + Al_2O_3$. Earlier we also noted such reduction of fracture toughness while studying ceramics similar to the material under investigation.

Thus, the tests in air show that silicon nitride ceramics exhibit degradation of thermomechanical properties below 1000° C. From the stress–strain diagrams for the material under study (Fig. 6.20) one can see that plastic strain occurs at higher temperatures, hence the softening of an intergranular phase traditionally used to explain the mechanical behaviour of ceramic materials at high temperatures is not responsible for their strength degradation. The above results as well as the stability of strength during the tests in a neutral medium lend evidence to the influence of oxidation on mechanical properties.

A SEM study of the specimens tested at 800° C in air revealed numerous cracks around the inclusions of oxidation products in the subsurface layer at

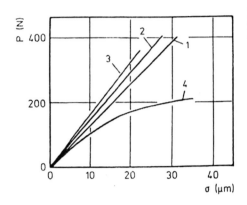

Fig. 6.20. Stress-strain diagrams of the NKKKM–84 ceramics at 20°(1), 800°(2), 1100°(3), and 1200° C (4)

a depth of 20–40 μm (Fig. 6.21). It should be noted that similar cracks were also formed in HPSN specimens with Y_2O_3 [6.5]. Such subsurface and surface cracks were also observed in a solid oxide layer formed on the specimens of silicon nitride ceramics prepared by different methods and containing various additives (Sect. 2.1). Such cracks could develop as a result of the densification of the surface layer due to the volume increase at the $Si_3N_4 - SiO_2$ transformation and the rise of internal stresses. Above 1000° C, with plastic effects appearing in the oxide layer and the intergranular phase, these stresses relax easily, and in the temperature range of interest brittle fracture of fragments of an oxidized surface layer can occur. However, the cracks observed could have developed during the cooling of the specimen rather than during the exposure at 800° C. Formation of the cracks along the boundaries of SiO_2 fragments, which have a higher thermal expansion coefficient, supports this assumption (on cooling, the shrinkage of SiO_2 regions exceeds that of Si_3N_4, and this results in spalling and cracking along their boundaries). The data of Table 6.1 also confirm that the cracks seen in Fig. 6.21 are apparently not an immediate reason for the strength degradation of the ceramics at 800° C: below 800° C the oxidation of the unloaded specimens does not result in their lower residual strength. At

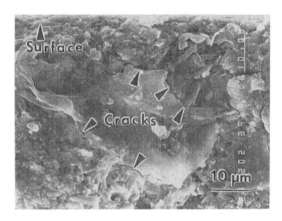

Fig. 6.21. Micrograph of a characteristic fracture with oxidation products

the same time it is possible that the formation of a brittle oxide phase of low strength can promote the propagation of a fracture crack, and the formation of this phase at the crack tip accompanied by a considerable volume increase (by more than 70%) can exert a wedge effect.

The above facts and arguments demonstrate quite definitely that the strength degradation at 800° C results from a simultaneous effect of mechanical stresses and chemically active media, i.e. the SCC of the material.

To confirm a close connection between chemical interactions in the material and its failure, the quantity of the oxidation products formed during the exposure of the loaded and unloaded specimens at 800° C in air was determined. It was found that this quantity increased with the growth of preload and loading time. This is supported by the direct measurements of the mass gain Δm of the specimens after their preloading, as well as by the results of XRD diffraction based on the determination of the relative intensity of α-cristobalite peaks (Fig. 6.22). It is interesting to note that the intensity of α-cristobalite peaks in the compression zone appeared to be even somewhat higher than in the tension zone. Preliminary compression tests [6.57] have unambiguously demonstrated the effect of such loading on the oxidation process (Fig. 6.23). The mass gain of the specimens increases with the increase of compression load (Fig. 6.24).

Charles and *Hillig* [6.36] have demonstrated that the rate V of interaction of ceramics and glass with H_2O is exponentially dependent on stress

$$V \sim \exp(k\sigma) \quad .$$

Other equations were also proposed [6.18]:

$$V = V_\infty \exp(-A/\sigma) \quad ,$$

and the power equation

$$V = A_1 \sigma^n \quad .$$

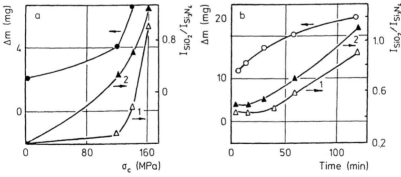

Fig. 6.22. Mass gain and relative intensity of α-cristobalite peaks as a function of loading for 30 min (a) and as a function of loading time under 140 MPa (b): tension zone (1); compression zone (2)

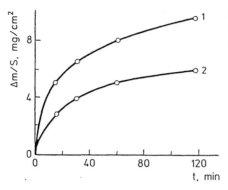

Fig. 6.23. Oxidation kinetics of NKKKM–84 specimens at 800° C on heating in air under a 120 MPa load (1) and unloaded (2)

Evans proved that the Charles and Hillig theory can successfully be applied to the description of a high-temperature subcritical crack growth and strength reduction of glasses and ceramics. This is in agreement with our data. The curves shown in Fig. 6.22a can be described by the exponential equation, though a small number of points complicates a exact analysis of the curves [6.58].

The performed estimations have demonstrated that at 800° C the increase of stresses from 230 to 690 MPa results in the rate constant k_p by an order of magnitude greater calculated by the equation

$$(\Delta m/m)^2 = k_p t + A \quad ,$$

where $\Delta m/m$ is the relative mass gain, t is the time (in our case it is 2 h), A is the constant. The apparent activation energy of Si_3N_4 oxidation calculated by the Arrhenius equation increases with temperature in the range of 700°–800° C. The results are in agreement with the data presented in [6.59], viz. when an external compression load is applied, the activation energy of a bond break increases, i.e. in this case the activation energy of oxidation E can be given as

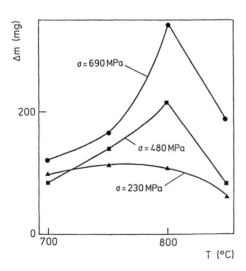

Fig. 6.24. Mass gain after compression tests during 3 h in air vs temperature curves of RBSN

$$E = E_0 + \Delta E(\sigma) \quad ,$$

where E_0 is the activation energy of oxidation in the unstressed state, $\Delta E(\sigma)$ is the variation of the activation energy as a result of external forces.

The very limited number of published data on the effect of stresses on Si_3N_4 oxidation [6.5, 8] does not allow to conclude whether there are any changes of phase composition and structure of the oxide layer together with an increase in the amount of oxidation products.

It should also be noted that the oxidation in the temperature range of interest can lead to strong internal stresses which give rise to failure (Fig. 2.49) or noticeable buckling of specimens due to nonuniform oxidation of the sides.

Taking into account the revealed effect of stresses on the oxidation rate of silicon nitride ceramics, one may suppose a possible self-accelerating oxidation process (similar to autocatalytic processes) proceeding in the range of intermediate temperatures. Such a mechanism of oxidation can explain the spontaneous acceleration of the process even without cracking of the specimens (Fig. 2.24).

Similar stress-enhanced or stress-assisted oxidation below 1000° C resulting in the loss of strength was also observed for denser high-strength HPSN with different additives [6.30] and SRBSN [6.60]. In all appearances, stress-assisted oxidation also exerts a considerable influence on subcritical crack growth in ceramics [6.61–63] determining, along with viscous flow of the grain-boundary phase, the sensitivity of HPSN to loading rates at high temperatures (Table 6.1).

As is shown in [6.64], oxidation caused failure of HS–110 and HS–130 HPSN specimens (Norton Co., USA) during cyclic fatigue tests above 1200° C. The HS–130 specimens with their higher purity and oxidation resistance were more durable. The authors of [6.65] also think that RBSN specimens during cyclic fatigue tests on tension at a 50 Hz frequency and 1200° C fail due to oxidation of silicon inclusions but not Si_3N_4. Oxidation is of considerable importance for thermocyclic tests of RBSN [6.48]. But in this case not only oxidation but also defects of the initial materials, the state of the grain-boundary phase and other factors are of great importance. Therefore, it is quite difficult to estimate the contribution of oxidation to the failure of ceramics during tests.

6.3.2 Silicon Carbide Ceramics

Above 1000° C time-dependent failure of HPSiC, sintered and self-bonded SiC materials was observed. Subcritical crack growth was most active in HPSiC [6.66]. Since its strength in vacuum is higher than in air [6.66], one may assume that oxidation promotes crack propagation. According to [6.67], the subcritical crack growth in HPSiC was proportional to the partial oxygen pressure. The failure of HPSiC at high temperatures is entirely intercrystalline. The grain boundaries are apparently in a higher energy state and are more susceptible to the effect of oxygen. The impurities and additives also accumulate along the grain boundaries. On this basis, *Henshall* [6.68] proposed a scheme of corrosion-mechanical crack growth.

At the same time the authors of [6.69] believe that a dominating mechanism inducing a slow crack growth in SiC is not oxidation but viscous flow of the grain-boundary phase.

A significant effect of oxidation has been observed in the cyclic fatigue behaviour of commercially available sintered and siliconized silicon carbides [6.47]. Beneficial strengthening effects after tension–tension cyclic loading (in bending at 5 cycles/h) at 1260° C in air were observed for sintered and siliconized silicon carbides when compared to strengths after static loading. These effects were not observed under cyclic loading in argon. Microscopic evidence suggests that the strengthening is due to oxide ligaments which bridge the crack surfaces during cyclic loading in air, thereby reducing the effective stress at the crack tip. Similar ligaments do not form during static loading because the cracks are held open. This effect is schematically illustrated in Fig. 6.25. In argon, the absence of oxidation products results in the behaviour similar to the static loading case. Results at higher frequencies tend to confirm this hypothesis, since the strengthening disappears, presumably at frequencies above the viscoelastic–elastic transition for the glass [6.47].

Self-bonded SiC containing free silicon is the most resistant to subcritical crack growth of all the silicon carbide materials (Fig. 6.26).

The results we obtained [6.70] allow to conclude that failure stresses in self-bonded SiC are insensitive to loading rates (in other words, the material can be stressed at high temperatures for a long time), and the change in mechanical behaviour is determined only by softening and, probably, by viscous flow of present silicon, which is one of major limiting factors for the operation of this material. The data on sintered SiC are rather contradictory [6.71.72].

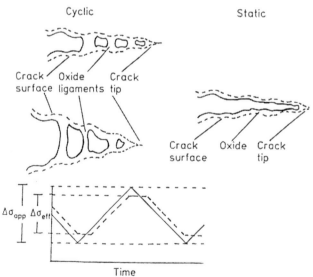

Fig. 6.25. Schematic representation of the crack tip in silicon carbide ceramics during static and cyclic loading in air at high temperatures [6.47]

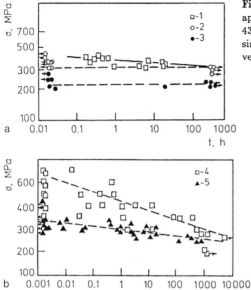

Fig. 6.26. Time to failure as a function of applied load for self-bonded SiC (a): NC–435(1); NC–433(2); Silcomp CC(3) and sintered SiC (b) at 1200° C. Year of development 1980(4), 1978(5) [6.46]

6.3.3 Alumina Ceramics

Alumina ceramics, just as glass, are affected by moisture [6.73]. The results of four-point bending tests for porous alumina are given in Table 6.3, with the highest values of strength being achieved in H_2O vapour-free liquid nitrogen.

Table 6.3. Strength of alumina ceramics in various environments

Environment at 22° C	Air	H_2O_{dist}	$N_{2\,liq}$
σ, MPa	414 ± 76	390 ± 62	489 ± 92

During the static fatigue tests in distilled water the specimens fail along the grain boundaries. The dissolution of impurities in water caused the intercrystalline failure of ceramics. The tests resulted in an increase of calcium and silicon contents in water, but the aluminium concentration did not change [6.73]. The authors of [6.32] studied several Al_2O_3-based materials with different purity in air, water and salt solutions, and also came to the conclusion that higher contents of impurities decreased the SCC resistance of ceramics.

6.3.4 Zirconia Ceramics

It has been reported that under the appropriate conditions transformation-toughened materials are subject to the detrimental effect of slow crack growth and low-temperature aging [6.74–77]. Extended exposure of toughened ceramics at 150°–300° C in the presence of water has been reported to lead to the premature $t \rightarrow m$ transformation and changes in mechanical strength.

The aging studies reported in the literature focused on the exposure of Y_2O_3-stabilized ZrO_2 at temperatures from 150° to 450° C in air for 25 to 1000 h. The degree to which degradation occurred depended on the amount of stabilizer, the particle size, aging time and the temperature to which the materials were exposed. Degradation increased as the concentration of Y_2O_3 decreased and the particle size of the material increased.

The dynamic fatigue analysis indicates a substantial slow crack growth for the PSZ and TZP when subjected to stress at both 25° and 250° C in moisture-containing environments. The slow crack growth observed in dry N_2 environments at 25° C for the PSZ material may be due to the presence of small quantities of water trapped at the crack tip or to some other mechanism not associated with water corrosion [6.74]. Zirconia ceramics of the conventional PSZ-type suffer a loss of strength after a long-term treatment at temperatures above 900° C. The reason for this thermal deterioration is the well known phase transformation and not the environmental effects [6.78]. It is believed that the effect of aging on strength is related to the thickness of the surface transformation zone. A surface transformation layer less than 100 μm results in compression surface stresses which enhance strength. However, if the surface transformation layer exceeds 100 μm it results in the generation of microcracks which reduce strength [6.74].

6.3.5 Salt-Assisted Strength Degradation

As seen in Fig. 6.17, an air medium gave rise to a local minimum of strength at $\sim$ 800° C on the σ vs T curves. Sodium salts brought about a still greater strength degradation [6.79], which reached its utmost when the materials were impregnated with the eutectic mixture (Fig. 6.12,14). It is substantial that the strength drop in the presence of the sodium salts was more pronounced in air than in argon.

The sodium salts encouraging a more extensive oxidation [6.80] lead to a decrease of the grain-boundary strength. It is natural to suggest that the kinetics of the process, as in the above-mentioned case of oxidation in the absence of salts, are influenced, along with temperature, by the amount of oxidation products and their state (solid phase or melt). Therefore it may be anticipated that the character of oxidation and its effect on strength in various temperature ranges will be different in particular features but similar in quality both in the presence and absence of salts. The data presented in Figs. 6.27, 28 also suggest that stress corrosion is most probable to occur at 800° C. Clearly, its activity will be hampered at high temperatures by the formation of a sufficiently thick continuous oxide film which was observed, in particular, after the tests at 1100° C (Fig. 6.28).

After impregnation with the salts the material degradation under static load is even more pronounced (Fig. 6.27) in the transition temperature interval than in air (Fig. 6.18). For instance, the long-term strength of RBSN impregnated

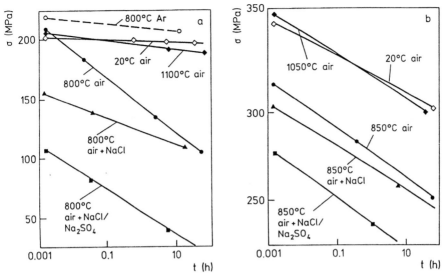

Fig. 6.27. Static fatigue diagrams for reaction-bonded NKKKM ceramics (a) and HPSN with TiN (b)

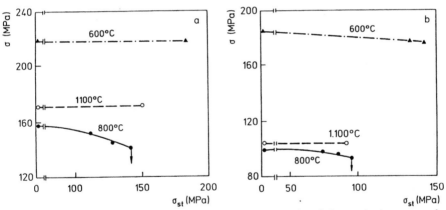

Fig. 6.28. The effect of a static load on the strength of reaction-bonded ceramics impregnated with NaCl (a) and NaCl/Na$_2$SO$_4$ (b)

with NaCl and the eutectic NaCl/Na$_2$SO$_4$ was 55% and 35%, respectively, of the material strength determined in the absence of aggressive media. The salts affected the mechanical behaviour of the hot-pressed ceramics in a similar manner (Fig. 6.27b). But no effect of salts on the slope of the static fatigue diagrams was observed.

As far as the effect of load and time on the material degradation in the presence of sodium salts is concerned, it should be emphasized once more that the specific mechanical properties at 800° C are distinct from those character-

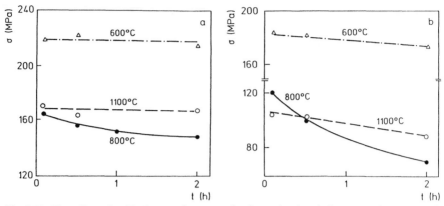

Fig. 6.29. The effect of oxidation on the strength of reaction-bonded ceramics impregnated with NaCl (a) and NaCl/Na$_2$SO$_4$ (b)

istic of the other intervals. As is seen in Fig. 6.29 illustrating the behaviour of the reaction-bonded NKKKM ceramics, the duration of the specimen oxidation under isothermal conditions in air prior to mechanical tests at 600° and 1100° C does almost not affect its strength. At the same time at 800° C a more prolonged oxidation progressively impairs the strength of the material impregnated with NaCl and especially with NaCl/Na$_2$SO$_4$. A similar phenomenon was observed when determining the long-term strength of preoxidized specimens of the same ceramics in air at the same temperature under a load of 66%, 76% and 84% of the median strength at a given temperature for 0.5 h (Fig. 6.28). At 800° C the strength decreases under increasing load so that under a load of 84% of the median strength the specimens could not bear the load for half an hour under isothermal conditions. In contrast, at 600° and 1100° C the same loads had almost no effect on the specimen strength.

A comparison of the test results in air and a neutral medium allows to attribute the weakening observable at 800° C to the presence of oxygen and the resulting oxidation of silicon nitride. When discussing the strength degradation mechanism for reaction-bonded ceramics in the absence of salts, it was shown in Chap. 2 that the temperature range under consideration corresponded to the onset of the material oxidation. As pointed out in the above-mentioned chapter, the specificity of oxidation at 800° C lies in the formation of a solid oxide layer consisting primarily of α-cristobalite and a glassy phase. These products do not cover the specimen surface with a continuous layer and do not involve a retardation of oxidation in contrast to higher temperatures when they melt to form a continuous oxide film preventing from further oxidation. This concept is obviously supported by the nature of the kinetic curves of oxidation according to which the mass gain of the specimen at 800° C increases monotonously with time, whereas at temperatures above 1200° C it reaches a saturation level within a short time interval (Fig. 2.15). This is in good agreement with the results of testing under static loading as the static fatigue diagrams of the long-term strength at 800° C in air have a greater angle of slope than those in argon or in air at 20° and 1100° C (Fig. 6.27). What was said above concerning reaction-bonded ceramics was also observed in the case of hot-pressed ceramics and therefore need not be repeated here. This allows to consider the above phenomenon as typical of most silicon-nitride–based ceramics.

The test for crack velocity vs stress intensity at 1300° C has shown that HPSN subjected to fracture in tensile loading exhibited slow crack growth in the presence of sodium sulfat, twice the rate was established when tested in air [6.81]. The results of Si_3N_4 (NC–132) during stepped-temperature stress-rupture (STSR) tests also indicate the importance of the superposition of environmental parameters such as temperature, medium and stress, in determining the behaviour of a material [6.82]. STSR results indicate that Na_2SO_4 additions have an adverse effect on the high temperature reliability of Y–TZP. This is noticeable at 900°–1000° C under stresses of 200–300 MPa. However, Ce–TZP, alumina and even Si_3N_4 subjected to the same test conditions appear to be unaffected [6.82].

At the same time the presence of sodium sulfate on the surface of the post-sintered RBSN containing free silicon (SRBSN–Si) at a small strain rate results in an abrupt growth of deformability and a weak increase in strength (Table 6.4), with horizontal sections on stress-strain diagrams [6.83]. A flow plateau on the diagrams is indicative of creep activation under the given experimental conditions.

Creep experiments [6.83] confirmed the effect revealed at short loading times for SRBSN–Si specimens covered with Na_2SO_4. And at longer loading times SCC and strength decrease characteristic of NKKKM and HPSN-TiN under similar conditions were not observed. This phenomenon may be explained by the Ioffe effect (uniform dissolution of a defective surface layer in soda-silica glass being formed) or the diffusion of sodium ions along the grain boundaries into inner layers of the material resulting in higher plasticity of the grain-boundary phase and grain boundary sliding. In the latter case lower strength of the specimens could be expected. The absence of this effect may be due to free silicon present in the structure of the material. Viscous flow can cause crack retardation, as it was observed for self-bonded SiC [6.70].

Consequently, there is still a deficit in the comprehensive treatment and the complete understanding of these important long-term–high-temperature phenomena in engineering ceramics. A better understanding may help to improve the quality of materials in the future.

Table 6.4. Mechanical properties of SRBSN-Si

Temperature, ° C	Environment	Loading rate, mm/min	Strength, MPa	Ultimate strain, %	E, GPa
20	Air	0.5	506	0.201	252
20	Air	0.005	510	0.202	252
1300	Air	0.5	390	0.457	140
1300	Air	0.005	191	0.671	111
1300	Na_2SO_4	0.5	410	0.427	189
1300	Na_2SO_4	0.005	222	1.572	117

6.3.6 Triboxidation

Oxidation of ceramics operating in friction couples is a particular case of oxidation [6.84]. As is well known, friction activates many chemical processes. Recent investigations performed on different materials have given rise to a new field of science – tribochemistry. At the same time, tribochemical processes occurring on friction of ceramic components have not yet been studied thoroughly enough.

Friction of sialon and silicon carbide ceramics resulted in amorphous silicate products of wear [6.85]. Oxidation and interaction with metals during cutting are one of the reasons of wear of cutting tools made of silicon nitride ceramics [6.86] and oxide ceramics with TiC or TiN additions [6.87]. The investigation of dry friction of sialon ceramics containing 10%–70% ZrN against steel [6.88] revealed pronounced oxidation of zirconium nitride. The fact that zirconia appears not only on the sliding surface but also on neighbouring lateral faces of the specimen (Fig. 6.30a) suggests that in this case the main reason of oxidation was the heating-up of specimens on sliding and not the activation of chemical processes due to mechanical energy. As a result, a 3 mm wide oxidized fringe was formed on the lateral face of the ceramic specimen adjacent to the sliding surface. XRD analysis demonstrated that the oxide layer included zirconium oxynitride ZrN_xO_y (shift of ZrN diffraction lines) and cubic and monoclinic zirconia modifications. As was shown in Sect. 2.6.1, the oxidation of zirconium nitride present in the composition of ceramics leads to considerable internal stresses. This probably caused the cracking of specimens containing over 30% ZrN during friction tests [6.88].

The friction tests of hot-pressed Si_3N_4, SiC, AlN and sialon ceramics in sliding contact with steel at a velocity of 1–15 m/s and loads of up to 6 MPa could not reveal corresponding oxides either on the surface of ceramics or in the wear products [6.89, 90].

At the same time, hot-pressed B_4C in sliding contact with steel [6.91] was oxidized under similar conditions. The AES analysis of the ceramic surface revealed Fe, O, C and B on the wear track. Thus, tribochemical reactions on

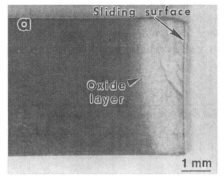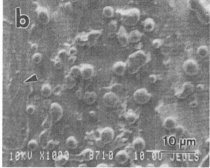

Fig. 6.30. Specimen of sialon with 50% ZrN after friction tests (a) and oxidation products on the sliding surface of hot-pressed B_4C (b)

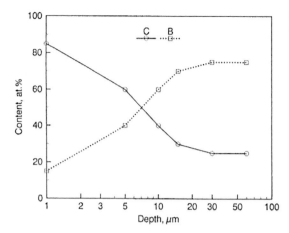

Fig. 6.31. Elemental depth profiles of the B_4C specimen after friction

the sliding surfaces result in the formation of a layer consisting of compounds in the system Fe–C–O–B. At the same time in some grooves of the wear track on the surface of ceramics a cracking glassy film was revealed (indicated by an arrow in Fig. 6.30b), the morphology of which was similar to that of the film formed during the oxidation of B_4C in air (Fig. 2.39). Near the grooves one can also see (Fig. 6.30b) solidified drops of 2–5 μm in size. The analysis demonstrated that the main components of these drops were boron, oxygen and iron. Thus, they can be classified as iron borates or their mixture with B_2O_3. The above confirms the oxidation of iron and boron carbide during the friction experiment. As is known, the temperature of initial oxidation of B_4C is close to 600° C. Under test conditions this process starts at lower temperatures, which is probably due to mechano-chemical activation. However, the formation of drops is indicative of the presence of a liquid oxide layer on the specimen surface. Taking into account that the melting temperature of B_2O_3 is $\sim$ 450° C, it remains to be assumed that the contact temperature was not less than 450°-500° C. The oxidation of B_4C in sliding contact with B_4C was also revealed in high-temperature friction tests [6.92, 93].

The analysis of fractures of the boron carbide specimens after the tests demonstrated that the tribochemical processes in their surface layer resulted in the formation of graphite and the removal of boron (Fig. 6.31), as it occurs during oxidation in air at much higher temperatures (Sect. 2.4).

6.4 Lifetime Prediction in View of Environmental Effects

Long-term loading of ceramic components often results in their failure due to subcritical crack growth. Thus, with known characteristics of resistance to subcritical crack growth, one can estimate the time for a crack to grow to a critical size. Therefore, the data on a flaw size and existing stresses make it possible to predict the lifetime of the material. Such estimation of service life for ceramic materials is rather interesting, since it does not require long-term tests and can be performed even at the stage of development.

To predict strength parameters, one can use the stress-intensity-factor–crack-velocity curves ($K_1 - V$ diagrams) (Fig. 6.19). The time to failure under static loading can be determined analytically in a generalized form, the time to failure (or failure stress) under cyclic loading can be determined by numerical methods or analytically with corresponding $V - K_1$ relations. The previous sections posed the problems of corrosion effect on the lifetime of ceramics. The methods for calculating the time to failure accounting for environmental effects are given below in more detail.

The basic framework for understanding the strength and toughness of ceramics is provided by fracture mechanics. The stress intensity factor of ceramics K_1 can be understood in terms of the Griffith equation

$$K_1 = \sigma Y \sqrt{c}, \tag{6.1}$$

where Y is a geometrical constant and c is a flaw size. Supposing that the time to failure consists entirely of the time required for the subcritical crack growth of a preexisting flaw and differentiating the equation (6.1) with respect to time under a constant stress, we get

$$\frac{dK_1}{dt} = (\frac{\sigma Y}{2\sqrt{c}})\frac{dc}{dt} + \sigma \sqrt{c}\frac{dY}{dt} \quad .$$

Taking into account that $V = dc/dt$, after transformation we get

$$dt = (2/Y^2\sigma^2)(K_1/V)(dK_1 - K_1 d\ln Y) \quad . \tag{6.2}$$

Since the variation of Y exerts a small influence on the crack growth equation,(6.2) can be reduced to the form $dt = (2/Y_0^2\sigma^2)(K_1/V)dK_1$, where Y_0 is the value of Y for the initial crack. Further integration gives

$$t_f = (2/Y_0^2\sigma^2) \int_{K_{1j}}^{K_{1c}} (K_1/V)\, dK_1 \quad .$$

This is a general form of expression for time to failure. Above a threshold stress intensity factor the crack velocity is proportional to the nth power of the stress intensity factor.

$$V = \alpha K_1^n \quad , \tag{6.3}$$

where α and n are the constants. Subject to (6.3) we get

$$t_f = [2/Y_0^2\sigma^2\alpha(n - 2)](1/K_{1j}^{n-2} - 1/K_{1c}^{n-2}) \quad .$$

Thus, the value of t_f depends on K_{1j} which is determined by the size of earlier existed defects and by stress [6.94]. A similar equation can be used, e.g., to

144

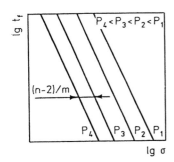

Fig. 6.32. Service-life–stress–probability diagrams based on the Weibull function for approximation of experimental data (slope of the curves is determined by the value of n) [6.94]

predict the service life of components by the data obtained on double torsion [6.94].

Weibull statistics are widely applied to the strength of ceramics, and for many data there is good agreement between experimental results and this function. If we use the Weibull distribution $P_s = \exp(-(\sigma/\sigma_0)^m)$, where P_s is the survival probability, σ_0 is the normalization factor, m is the homogeneity coefficient (usually known as the Weibull modulus), to describe the results of fracture tests. Then the time to failure at a failure probability of $P < 0.1$ takes the form

$$\lg t_f = (n - 2/m)\lg P - n\lg \sigma + \lg[2/\alpha(n-2)Y^2 K_{1c}^{n-2}] - (J/m) \quad ,$$

where J is the constant determined by the shape of specimens and the concentration of defects. This result can be presented as stress–service-life diagrams (Fig. 6.32) which are used to estimate the applicability of the material under any conditions when maximum stresses and requirements to service life and permissible failure probabilities are known.

Such an approach to the description of subcritical crack growth in ceramics supposes the power $V - K_1$ relation (6.3). Alternative descriptions of subcritical crack growth can take an exponential form. *Charles* and *Hillig* [6.36] proposed a quantitative theory based on the concept of a stress-dependent corrosion reaction. This theory and other theories accounting for thermally and chemically activated crack growths give the velocity of chemically assisted crack growth as an exponential function of the stress intensity

$$V = \alpha_1 \exp[n_1(K_1/K_{1c})] \quad .$$

According to the data of [6.95], this equation is more suitable than the power equation for describing the behaviour of HPSN at 1200° C under a static load, linear cyclic loading (linearly varying stress from zero to maximum, frequency is 7 cycles per minute) or loading with a constant stress growth rate. For the first version of loading the following equation was derived [6.95]:

$$t_f = \left(\frac{2K_{1c}^2}{\sigma^2 Y^2 \alpha_1^2 n_1}\right) \exp\left(-n_1 \frac{\sigma}{S}\right)\left(\frac{\sigma}{S} + \frac{1}{n_1}\right) \quad , \tag{6.4}$$

and for the second and third ones

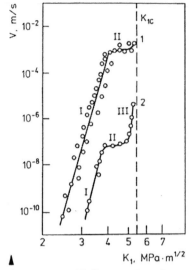
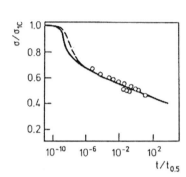

Fig. 6.33. $K_1 - V$ diagrams for polycrystalline alumina tested in air of 50% humidity (1) and in toluene (2) [6.96]

Fig. 6.34. Universal fatigue curves for polycrystalline Al_2O_3 in humid air (solid line) and in toluene (dashed line). The experimental data are shown with dots. $t_{0.5}$ is the time to failure under a stress equal to half the failure stress in an inert medium (σ_{1c}) [6.96]

$$t_f = \left(\frac{2K_{1c}^2}{\sigma^2 Y^2 \alpha_2 n_2} \right) \exp\left(-n_2 \frac{\sigma}{S} \right) \quad , \tag{6.5}$$

where S is the failure stress in the absence of subcritical crack growth. According to the data of [6.95], (6.4, 5) are used to predict the time to failure for HPSN at high temperatures.

Since the form of the $K_1 - V$ diagram is dependent on the mechanism of crack growth and varies with the change of the corrosion effect of the environment (Fig. 6.33), the prediction of the service life of the material by the $K_1 - V$ diagrams accounts for corrosion effects [6.96].

Figure 6.34 presents the fatigue curves obtained from the calculations by the $V - K_1$ diagrams (Fig. 6.33) and experimentally [6.96]. The section where the curves for toluene and air tests are quite different corresponds to section II of the $K_1 - V$ diagram, while the section where the curves coincide corresponds to section I of this diagram.

Such a description will be exact enough if the subcritical crack growth is caused by the interaction with the environment. If corrosion exerts a greater influence on the initiation of defects, but not on crack propagation, the above approach is not very suitable. Several equations to calculate the time to failure under a load considering environmental effects were given earlier.

For failure under variable stresses, by integrating (6.1) and taking into account (6.3), we have

$$V = \frac{dc}{dt} = \alpha \sigma^n Y^n c^{n/2}$$

$$\int_0^{t_f} \sigma^n dt = \frac{1}{Y^n \alpha} \int_{c_0}^{c_c} \frac{dc}{c^{n-2}} = \frac{2}{(n-2)Y^n \alpha} \left(\frac{1}{c_0^{n/2-1}} - \frac{1}{c_c^{n/2-1}} \right) \quad ,$$

where c_0 and c_c are initial and critical flaw sizes. The substitution of the stress–time relation in the equation and further integration of the left part of the equation allow to calculate the time to failure or failure stress from the initial crack length c_0. Thus, at a constant stress growth rate $\dot{\sigma}$

$$\sigma_f^{n+1} = \frac{2(n+1)}{\alpha Y^2 (n-2)} \dot{\sigma} \left(\frac{S}{K_{1c}} \right)^{n-2} \quad .$$

For cyclic loading the integration is much more complicated [6.94].

Thus, the $K_1 - V$ diagrams can be used to predict the mechanical behaviour of ceramics without long-term fatigue tests. However, to get actual data, these diagrams should be plotted for the environments corresponding to the service conditions of a given material. For example, thermal fatigue behaviour is predicted from the data on the temperature dependence of crack growth rate:

$$V = \alpha \cdot K_1^n \exp(-Q/RT) \quad , \tag{6.6}$$

where Q is the activation energy of crack growth. In [6.97] for silicon nitride ceramics at $1100°$–$1400°$ C the value of $n = 6$ and $Q = 715\,\text{kJ/mole}$ were calculated. Knowing the values of K_{1c}, σ, flaw sizes and the equation describing the crack growth rate as well as the parameters characterizing the variation of thermal and, thus, stressed state of a component during thermocycling, one can compute the crack growth during one heating cycle. So it allows to determine the number of cycles to failure at a certain temperature difference or the critical temperature difference (necessary for failure during one "heating–cooling" cycle). The calculations are in satisfactory agreement with the results of thermal fatigue tests [6.98]. However, the values of α, Q and n in equation (6.6) are valid only for specific temperature ranges, materials and environments. Therefore, for other conditions a subcritical crack growth should be investigated experimentally to derive the equation describing the rate–temperature relation.

The development of valid prediction methods is of great importance for reliably determining the service life of ceramic components [6.95, 99, 100].

Since the analytical estimation of lifetime under complex cyclic loading occurring in operation of components of engines and other devices is difficult, long-term static and cyclic fatigue tests are used in many cases [6.46, 64].

The other reason for long-term fatigue tests is that the existing prediction methods do not always give correct results. All of them are based on the concept that a time-dependent failure of materials is caused by subcritical crack growth. At the same time, as it was shown in [6.34], though the failure of silicon nitride ceramics is influenced by moisture at room temperature and temperatures close to it, it is not associated with subcritical crack growth.

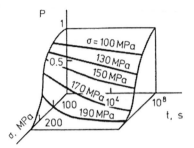

Fig. 6.35. Survival probability under long-term loading as a function of applied load and time for NKKKM ceramics [6.101]

In [6.102] for reaction-bonded silicon nitride-based NKKKM ceramics (Sect. 2.1), using the probabilistic approach, the equation for the survival probability P_s under long-term loading was derived:

$$P_s = \exp\left[-\left(\frac{\sigma}{\sigma_0}\right)^m \cdot \left(\frac{t}{t_0}\right)^{m/n}\right] \quad , \tag{6.7}$$

where σ_0 and t_0 are the normalization factors. This equation predicts the probability of failure for a preset test period (t) and stress (σ). The distribution of survival probability under long-term loading in a wide temperature range for NKKKM ceramics is given in Fig. 6.35 obtained from the calculations by the $K_1 - V$ diagrams (Fig. 6.36). The predicted survival probability decreases with an increase in loading time, but the extent of the decrease of P_s depends on the applied load. The comparison of experimental data on the life of the material with the calculations by equation (6.7) for room temperature is presented in Fig. 6.37. Here the slope of the experimental curve (and, thus, the homogeneity coefficient m/n) is close to the slope of the calculated curves. But the calculated distribution of survival probability corresponding to a stress of 170 MPa gives a somewhat undervalued P_s compared with experimental data which fall

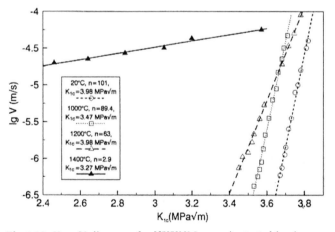

Fig. 6.36. $K_1 - V$ diagrams for NKKKM ceramics tested in air

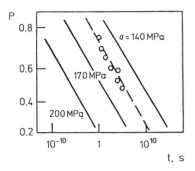

Fig. 6.37. Calculated and experimental survival probability vs loading time curves. Experimental lifetime values at $\sigma = 172$ MPa are shown with dots

on the calculated curve for $\sigma \approx 155$ MPa. This discrepancy can be explained by the fact that subcritical crack growth was not the only mechanism of ceramics failure or by errors in double torsion tests used for obtaining $K_1 - V$ diagrams (Fig. 6.36).

It is important to solve also the inverse problem – to predict an exponent n value and compare it with experimental results. Equation (6.7) can be written for this purpose as [6.102]

$$P_s = \exp\left(-\frac{t}{t_0}\right)^{m_t} \quad , \tag{6.8}$$

where

$$m_t = m/n \tag{6.9}$$

and

$$t_0^{eq} = t_0(\frac{\sigma}{\sigma_0})^n \tag{6.10}$$

Thus, there are two different opportunities for computating of n: from m and m_t (6.9) using survival probability distributions, and from t_0^{eq} at the median lifetime. For the latter case the n values can be obtained from tests at two different applied stresses σ_1 and σ_2. Also, n can be defined by

$$n = \frac{\lg(t_1/t_2)}{\lg(\sigma_2/\sigma_1)}, \tag{6.11}$$

where t_1 and t_2 are the lifetimes at σ_1 and σ_2 stresses.

The analysis of the data for NKKKM ceramics (Fig. 6.18, 19) based on the power law of subcritical crack growth (6.3) and equation (6.7) is shown in Fig. 6.38 and Table 6.5. As calculated by (6.9, 11) and measured at double torsion tests, activation of a subcritical crack growth at 800° C occurs. The absence of an accordance between measured and calculated n values (Table 6.5) can be explained by the differences in the propagation features of macrocracks at double torsion experiments (Fig. 6.19) and microcracks at bending tests (Fig. 6.38).

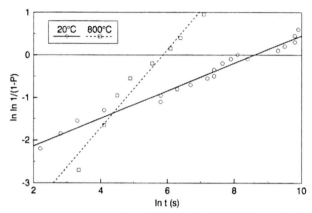

Fig. 6.38. Probability vs time diagrams for NKKKM ceramics tested in air

Table 6.5. Measured and predicted n values for NKKKM ceramics tested in air

The method for the estimation of n	Double torsion	By equation (6.9)	By equation (6.11)
n values at:			
20° C	146	74	75
800° C	62	27	35

We presented above only some equations for mechanical behaviour prediction applied to structural ceramics by different researchers. Quite a number of approaches to lifetime prediction of ceramics are described in more detail, e.g. in [6.103, 104]. These and other publications provide the necessary relations, sufficient for lifetime prediction of ceramics at least at the first stages.

7. Corrosion Protection and Development of Corrosion-Resistant Ceramics

The coating of a ceramic surface with nitrides, carbides or oxides is a key method of corrosion protection. As it was already mentioned in Chap. 1, chemical-vapour deposited (CVD) silicon nitride and silicon carbide exhibit a very high oxidation resistance. Oxide coatings can also reliably protect from oxidation. To enhance the corrosion resistance of materials in molten metals, their surface is coated with aluminium nitride. An optimum performance is achieved by coating a base material with the same compound, i.e. Si_3N_4 on silicon nitride ceramics, SiC on silicon carbide ceramics and so on. In this case the coating adheres well to the base material, the risk of its delamination or cracking on heating because of the difference in the thermal expansion coefficients of the coating and the base material is reduced to a minimum. The principal methods of coating are CVD [7.1] and plasma spraying [7.2]. For porous materials the impregnation with different compositions which creates a dense corrosion-resistant surface layer is also used. In the following we describe the protection methods in more detail.

7.1 Chemical-Vapour Deposited Coatings

Protective silicon nitride coatings are usually formed as a result of the reaction between silicon halides (SiF_4, $SiCl_4$ and others) or silane SiH_4 and ammonia [7.3].

By controlling the deposition parameters (temperature, vapour pressure, consumption and ratio of chemical agents, deposition chamber volume, etc.), one can vary the properties of Si_3N_4 deposits over wide ranges. The mechanisms of Si_3N_4 CVD are discussed in several publications [7.3–5]. The description of CVD reactors and the analysis of different methods of coating synthesis are given in [7.4]. Therefore, our examination of Si_3N_4 coatings will be confined mainly to the problems of corrosion protection. Thus, fine-grained equiaxed-crystal coating can be produced by nucleation-controlled vapour deposition [7.6]. The comparison of experimental values of the activation energy of the process (66.9 kJ/mole) with the thermal effect of the exothermal reaction $3\,SiCl_4 + 4\,NH_3 = Si_3N_4 + 12\,HCl$ (225.7 kJ/mole) shows that the Si_3N_4 formation proceeds by a complex mechanism with the formation of intermediate products, e.g. imides.

Table 7.1. Characterization of coatings

Coating	T_{dep}, °C	Structure	Colour	γ, kg/m³
C	1400	Crystalline (α-Si$_3$N$_4$) Plane $\langle 222 \rangle$ parallel to the base surface	Black	3180
A(I)	1300	Amorphous	White	2950
A(II)	1200	Amorphous	White	2900

The oxidation resistance of RBSN specimens coated with 0.05 mm Si$_3$N$_4$ having grain sizes of 1–10 μm becomes two orders higher on cyclic heating up to 1200° C [7.6].

Oxidation of chemical-vapour (SiCl$_4$ + NH$_3$ + H$_2$) deposited Si$_3$N$_4$ coatings (Table 7.1) was investigated in dry oxygen at a pressure of 0.1 MPa [7.7].

Crystalline and amorphous Si$_3$N$_4$ exhibit high oxidation resistance up to 1600° C (Fig. 7.1) and, thus, create a reliable protection for ceramic products. Oxidation of amorphous Si$_3$N$_4$ at temperatures close to 1600° C results in its partial transition into α- and β-modifications. The only oxidation product revealed is α-cristobalite. The kinetic curves of oxidation of crystalline Si$_3$N$_4$ at 1550°–1650° C as well as amorphous A(I) at 1550°–1630° C and A(II) at 1580° C obey the parabolic law. The apparent activation energy of oxidation of materials C and A(I) in the above temperature ranges is 390 and 460 kJ/mole, respectively.

At the same time, the kinetic curves for A(I) at 1650° C and A(II) at 1600° C are linear (Fig. 7.1), and at these temperatures a porous oxide layer is formed on their surface. The authors [7.7] are of the opinion that the transition from parabolic to linear oxidation kinetics is explained by the Si$_3$N$_4$ crystallization occurring under these conditions. However, pores formed in the oxide layer may result from reaction (2.4). On oxidation of CVD Si$_3$N$_4$, its oxide layer can remain in a solid state up to very high temperatures, since it consists of pure silica, without any additives or impurities.

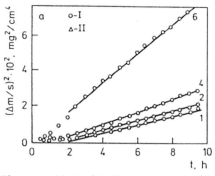 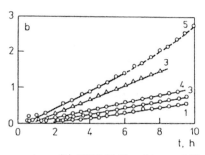

Fig. 7.1. Oxidation kinetics of crystalline (a) and amorphous (b) CVD Si$_3$N$_4$ of types A(I) – (I) and A(II) – (II) at 1550°(1), 1570°(2), 1580°(3), 1600°(4), 1630°(5), and 1650° C (6) [7.7]

It was also proposed to coat the surface of Si_3N_4-, BN-, AlN-, Al_2O_3-, SiO_2- and ZrO_2-based materials with sandwiches consisting of amorphous Si_3N_4 (deposition at 1000°–1200° C) and crystalline Si_3N_4 (deposition at 1200°–1500° C) layers. Such coatings exhibit high oxidation and thermal shock resistances.

For importing some other special properties apart from corrosion resistance (higher erosion resistance, hardness, etc.), Si_3N_4 − TiC sandwich coatings can be deposited by intermittent feed of nitrogen- and silicon-containing gas to a $TiCl_4$, H_2, CH_4 mixture. The thickness of each layer is < 1 μm, the number of layers is over eight, the total thickness of the coating 15 μm. One can also obtain amorphous Si_3N_4 matrix composite coatings containing TiN, BN and other reinforcing phases [7.8].

Silicon carbide coatings are produced by CVD of SiC from volatile silicon halides and hydrocarbons by reactions of the type $SiCl_4 + CH_4 = SiC + 4\,HCl$ or by thermal dissociation of gaseous organo-silicon compounds.

By controlling the parameters of a deposition process, one can vary the properties of the produced coatings [7.9]. Thus, at low temperatures fine-grained and metastable structures are formed. Crystal sizes grow with temperature. At 1400° C and low deposition rates single crystals and epitaxial layers of SiC are formed. An average size of crystals in a SiC layer deposited from trichloromethylsilane is 1 μm at 1400° C and 15 μm at 1800° C.

The growth rate of CVD silicon carbide does not exceed 0.5 mm/h. At the same time, comparatively low deposition temperatures (1100°–1550° C) make SiC coatings compatible practically with any structural material.

A major disadvantage of these coatings are residual stresses [7.10] arising because of the difference in thermal linear expansion coefficients of the coating and the base material (except for SiC deposited on SiC) and anisotropy of the coating. Due to comparatively low deposition temperatures, stresses do not relax and coatings crack. One of the methods to avoid this disadvantage are sandwich coatings with equally thick layers of pyrocarbon and SiC deposited from a chloromethylsilanemethane mixture. The sandwich coatings allow to control the thermal expansion coefficient and thermal conductivity and reduce arising stresses as a result [7.11].

CVD SiC coatings are used by Ultramet Co. (USA) for improving high-temperature corrosion resistance of various ceramic materials. Tubes made of porous recrystallized SiC with a CVD SiC coating operate in recuperators at 1000° C and exhibit a high thermal shock resistance [7.12]. Silicon carbide coatings deposited on the surface of RBSN by trichloromethylsilane decomposition at 920° C also gave positive results [7.13,14]. A material with 25% porosity coated with a SiC layer possesses nearly absolute creep resistance in air at 1200° C due to the absence of inner oxidation.

A silicon carbide coating of 0.025–0.5 mm thickness deposited on sintered and hot-pressed SiC increases their strength, wear resistance and oxidation resistance at 1300° C approximately by an order [7.6].

As is seen in Fig. 7.2, CVD SiC exhibits a high oxidation resistance even in wet O_2.

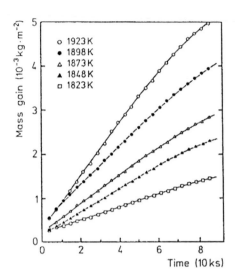

Fig. 7.2. Relationship between mass gain and oxidation time for CVD-SiC in wet O_2 [7.15]

CVD is one of the coating methods for titanium and zirconium diborides, other borides, carbides and nitrides. However, these coatings are less corrosion-resistant than the Si_3N_4 and SiC ones in most gas atmospheres and liquid media.

Although it is possible to form oxide coatings by chemical vapour deposition [7.16], more often they are produced by plasma spraying.

7.2 Sprayed and Sputtered Coatings

In practice coatings are often deposited by the plasma method [7.2] in which a coating material is fed to a high-temperature high-velocity plasma jet. The coating material in the plasma jet melts partially or totally, and on collision with the surface to be coated adheres to it. Though the coating material stays in the plasma jet for a relatively short time, its high temperature makes it possible to deposit the most high-melting coating materials. Argon, nitrogen and hydrogen are most often used as plasma-forming gases. Various designs of plasma spraying units are described in [7.17, 18].

Plasma spraying can be used to deposit any materials, except easily subliming ones, even on large-size components; the products to be coated are heated inconsiderably. However, this method has several disadvantages: It is energy-intensive and rather complex; process conditions providing coatings of uniform thickness are difficult to stabilize; it is quite difficult to deposit pore-free coatings, and shaped parts cannot be coated by this method.

The technology of flame spraying does not differ very much from that of plasma spraying. A coating material is fed to an incandescent gas stream of combustion products of corresponding mixtures. An acetylene/oxygen mixture is used most often. The temperature of the flame is much lower than in a plasma jet. This method can be used to deposit materials with a melting temperature up to 2500° C [7.19].

As far as energy intensity and complexity of process equipment are concerned, flame spraying is preferable to plasma spraying. The advantages of the flame method are the same as of the plasma method, except for the limitation on the melting temperature of deposited materials. It is quite difficult to control the gas composition of the flame, which also confines the range of deposited materials mainly to oxides.

To reduce the porosity of oxide coatings deposited by the above methods, it is proposed to impregnate them with intermetallic compounds of aluminium, zirconium, titanium, silicon, chromium or tantalum with further thermal treatment in air. The oxides of metals close the pores in the coating. Another method of densification of a deposited coating is a laser treatment of the surface [7.20]. The Al_2O_3, ZrO_2, Y_2O_3, CaO and HfO_2 coatings can be densified by this method, but after the fusion of coatings with a laser beam they reveal minute cracks. Only on TiO_2-coated specimens there appeared no cracks [7.20].

Plasma spraying is used to deposit ZrO_2 [7.21], Al_2O_3, TiO_2, MgO, Y_2O_3, CaO, SiO_2 coatings as well as coatings consisting of their compositions. The thickness of coatings reaches 2 mm.

Sometimes ceramic coatings are deposited by cathode sputtering [7.16] consisting of the bombardment of the cathode made of a sputtered material with ionized gas atoms. The bombardment results in atomic sputtering of the cathode material. Atoms are deposited on the specimen to be coated which is placed between cathode and anode, forming a film with time.

The advantages of cathode sputtering are: thin films can be produced from the most high-melting materials; good adhesion of films with the specimen; on sputtering of alloys the composition of the film duplicates the composition of the cathode.

However, cathode sputtering requires complex equipment and a low rate of film growing.

Glow-discharge cathode sputtering is used to deposit fine-grained and amorphous coatings up to 50 μm thickness [7.16].

Thin protective coatings are also deposited by electron beam evaporation [7.22]. In this case the coating material is evaporated by an electron beam. Ion or electron beam evaporation is used to deposit silicon nitride coatings up to a thickness of 10 μm.

In [7.23] oxidation of coated RBSN was studied in air at 1000°–1200° C. Polycrystalline silicon with a purity of 99.99% was used as a target. The 25 μm thick coating with a density close to the theoretical one was deposited on RBSN of 98% purity for 1–5 h at a gas pressure of 0.5–2 MPa and a voltage of 0.5–2.7 kV. For this purpose argon as an ionizing gas and nitrogen, oxygen and N_2O as reactive gases were used. The coated specimens were annealed in nitrogen with 1% hydrogen addition at 1300°–1500° C. Before coating the RBSN substrate was polished and ultrasonically cleaned in acetone, methanol and distilled water. Such a coating consisting of α- and β-Si_3N_4 as well as of unreacted

Table 7.2. Comparative effect of various methods for corrosion protection on the strength (MPa) of reaction-bonded NKKKM ceramics

Treatment	Testing temperature, °C		
	20	800	1000
Initial material	229	205	224
Coating			
Al_2O_3	221	234	248
Si_3N_4	225	230	248
Impregnation			
$Al(NO_3)_3$	220	268	230
$Zr(NO_3)_4$	230	240	248
ETS*–32	250	286	256
ETS–32+$AlCl_3$	236	220	238
Thermal treatment for 3 h at 1200° C in air	–	252	–

* Ethyl silicate

silicon protects ceramics from inner oxidation at low temperatures and retards the conversion of silicon nitride of the base material into silica at higher temperatures. A high-temperature oxidation rate is less dependent on the coating thickness, since above 1000° C a continuous oxide layer is formed on the specimen surface. To get qualitatively good coatings, one should properly choose the amount of silicon in a coating and thoroughly clean the surface of the material.

Detonation-deposited Al_2O_3 and Si_3N_4 coatings did really allow to avoid the reduction in strength at 700°–1000° C and to protect ceramics from stress corrosion (Table 7.2). Coatings up to a thickness of 50 μm adhered well to the RBSN substrate and did not delaminate under mechanical loading (Fig. 7.3a). Thick oxide coatings delaminated during high-temperature mechanical tests (Fig. 7.3b) due to the difference in thermal expansion coefficients and elastic properties of Al_2O_3 and Si_3N_4.

It has been shown in [7.24] that a plasma-sprayed mullite coating provides a promising protective layer for both heat exchanger tubes and heat engine components. Mullite matches SiC in its thermal expansion and the stoichiometric

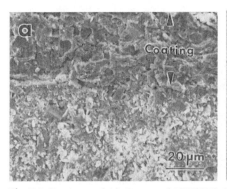

Fig. 7.3. Fractures of Al_2O_3-coated NKKKM–84 specimens

$2SiO_2 \cdot 3Al_2O_3$ compound with no excess SiO_2 shows good hot corrosion resistance. One difficulty with these coatings is their porosity. However, if thicker layers are applied and/or the porosity is closed off, minimal salt penetration occurs. The coated specimens show Na_2SO_4 deposition but no silicate formation or attack during 20 h exposure. The uncoated specimens exhibit the typical massive sodium silicate products discussed earlier on.

The investigation of corrosion resistance of coatings usually used to protect metals [7.25] has demonstrated that oxides of nickel, cobalt and magnesium are of limited use in highly contaminated atmospheres. All of these oxides react with SO_3 to form sulfates. A typical reaction might be

$$MeO + SO_3 \rightarrow MeSO_4 \quad .$$

The sulfates do not afford any protection to the underlying component and hence these oxides are apparently of limited use as thermal barriers.

Zirconia, however, is resistant to sulfidation and therefore would appear to be an excellent choice as a barrier layer in atmospheres which are contaminated with sulfur. The problem is that ZrO_2 undergoes a phase transformation at $1170°$ C from monoclinic to tetragonal unless it is stabilized. The stabilizing additives are usually MgO, CaO and Y_2O_3, and all of these are sulfated by reaction with SO_3. When vanadium is present in the fuel, a reaction can also occur with the stabilizing agents in the zirconia coating such as CaO, MgO or Y_2O_3. The formation of yttrium vanadate YVO_4 after exposure to vanadium-contaminated environments has been well documented [7.25]. The effect of such a reaction is a removal of the stabilizing phase and, as discussed earlier, this leads to an accelerated degradation of the coating because of the volume change associated with the unstablized material as it undergoes its phase transformation from the tetragonal to the monoclinic form.

It should be noted that all the test results referred to in this section have to be reported in terms of cycle life to cracking or spalling, not in terms of section loss or mass change. This is because the degradation of the ceramic coating system by the corroding media again is ultimately related to a mechanical incompatibility problem and therefore it is suggested in these systems that gravimetric techniques are of limited use.

7.3 Impregnation and Other Methods of Protection

The results obtained on impregnation of RBSN specimens of 28% porosity with liquid silicon in nitrogen were rather promising [7.26]. The surface of specimens after such treatment was covered with a dense protective layer of 40–$100\,\mu m$ thickness consisting only of β-Si_3N_4, which is quite resistant to oxidation. At the same time its content in the specimens did not exceed 4%. Mass gain for the specimens with a 45-μm coating during their oxidation up to $1400°$ C is an order lower than for unprotected specimens. With a 100-μm film, mass gain was not observed up to 1400° C, and the specimens exhibited high oxidation resistance in humid air at 1400° C, whereas unprotected specimens were 30% oxidized under similar conditions. Despite all the good properties of a silicon-impregnated material, it is rather difficult to apply this method of protection in practice, since the specimen surfaces should thoroughly be cleared from oxygen-containing silicon compounds, because SiO_2- or Si_2N_2O-containing surfaces are not wetted with liquid silicon.

Silicon carbide materials are proposed to be coated from a silicon slip and then to be treated in the stream of a 1:3 methane/hydrogen or ethane/hydrogen mixture at 1400° C. This treatment results in carbon deposition and SiC formation in the pores of the product. Carbon excess is removed by further treatment in air at 600° C [7.27].

The service life of porous materials based on Si_3N_4, SiC and other compounds can be increased by coating their surface with finely dispersed Si_3N_4 and Si_2N_2O from an aqueous slip followed by drying and firing above 1000° C [7.27].

According to the data of Norton Co., RBSN can be impregnated with the solutions of yttrium and aluminium nitrates, zirconium oxychloride or its organic compounds to protect it from oxidation. Multiple impregnation of the specimens is followed by their nitridation. Decomposition products formed as a result of the reactions

$$4\,Al(NO_3)_3 = 2\,Al_2O_3 + 12\,NO_2 + 3\,O_2 \quad ,$$

$$4\,Y(NO_3)_3 = 2\,Y_2O_3 + 12\,NO_2 + 3\,O_2$$

fill the pores on the specimen surfaces. In this case the oxidation resistance of the specimens increases without any changes of mechanical properties. But if the materials with the addition of zirconium compounds are exposed for 50 h in air at 1230° C, their strength increases considerably. A simultaneous increase in strength and oxidation resistance of reaction-bonded materials can be achieved by impregnation with a $ZrOCl_2$ and $Y(NO_3)_3$ mixture (8% Y_2O_3 after decomposition) and annealing in nitrogen for 8 h.

RBSN components can be impregnated with silicon halides followed by ammonia treatment. Here the processes are more or less the same as on the CVD of Si_3N_4, except that nitride forms mainly in the pores of the surface layer. Instead of silicon halides, these components can be impregnated with SiH_2NH or liquid organosilicon polymers [7.28]. Then they are heated in nitrogen or argon up to 500° C at a rate of 2° C/min and up to 1000° C at a rate of 10° C/min. The impregnation-pyrolysis cycle is repeated three times. The impregnation of specimens with organosilicon compounds (dimethylsiloxydiphenylsiloxycarborane and others) results in the formation of β-SiC in the pores of ceramics. On impregnation with SiH_2NH, α-Si_3N_4 and silicon appear [7.29]. The impregnation with organosilicon compounds followed by pyrolysis increases not only oxidation resistance but also hot corrosion resistance [7.30].

The impregnation of NKKKM ceramics with zirconium and aluminium nitrates or organosilicon compounds followed by thermal treatment (Table 7.2) imparts higher mechanical properties to ceramics and protect them from oxidation and stress corrosion over the range of intermediate temperatures.

Porous SSN ceramics can be protected from the effect of molten aluminium by impregnating with barium compounds (sulfate, chloride, hydroxide, stearate and others). Material containing pores filled with barium compounds, owing to a low thermal expansion coefficient, higher thermal shock resistance and poor wetting, operates well in molten aluminium and its alloys at 760° C [7.27].

The above processes are not always technologically easy to apply. A simpler method of protection is preoxidation of ceramic components under certain conditions. As it was already noted, the performance of the materials can be maintained under conditions favourable for a dense protective oxide layer to be formed on their surface (on operation in an oxidizing medium at high temperatures). However, over a certain temperature range (for most materials these are temperatures close to 1000° C) the oxidation rate is already high enough, but the protective layer does not form. This is particularly dangerous for porous materials, since it can result in their intensive inner oxidation. To eliminate undesirable phenomena, it is proposed to generate a dense oxide layer artificially by charging reaction-bonded materials into the furnace heated up to 1300°–1350° C for over 0.5 h [7.31].

After preoxidation of NKKKM specimens for 30 min at 1350° C, their further exposure in air at 1000° C for 10 h did not change their mass. Thus, the stability of ceramic properties can be improved in the process of operation.

According to the data of AiResearch Manufacturing Co. (USA), a 50 h exposure in air at 980° C also increases the strength of HPSN. Such treatment is used for turbine rotor blades manufactured by the company. Reaction-bonded materials are treated at 1400° C for 2 h or for 1 h at 1460° C followed by slow cooling together with the furnace down to room temperature.

Preoxidation under certain conditions can not only form a protective oxide layer on the specimen surfaces but also remove impurities from the material [7.32]. As it was shown in Chap. 2, additives and impurities with a high affinity for oxygen diffuse to the surface during oxidation of Si_3N_4 and SiC ceramics. Thus, the inner layers of the material are cleared from these substances. As is known, purer materials exhibit much higher oxidation resistance. Oxidation of dense Si_3N_4 materials with MgO addition at 1400° C for 300 h followed by the removal of the oxide layer improves their high-temperature strength by 50% and increases fracture toughness [7.32]. Reaction-bonded ceramics do not display a considerable increase in strength but their oxidation resistance becomes somewhat higher as a result of such treatment [7.33].

Silicon nitride ceramics can also be cleared from impurities by applying a strong constant electric field to the specimen at high temperatures. Here the impurities are removed not by the gradient of chemical potential as in the case of oxidation, but by the attraction of impurity cations to the cathode. The specimens are heated in an inert atmosphere up to 1100°–1400° C and exposed to the electric field with a $2 \cdot 10^4$ V/m intensity for 10–100 h. After cooling, the surface layer with the concentrated impurities is polished off [7.34].

A promising method of increasing corrosion resistance of materials is their surface treatment with high-energy ion beams for ion implantation or for amorphisation of a ceramic surface [7.35]. Covalent materials (α-SiC) acquire amorphism more easily than materials with ionic bonds (α-Al$_2$O$_3$).

7.4 Choice of Optimum Ceramic Composition

Corrosion protection of ceramics by coating, impregnation and preoxidation complicates the manufacturing process of ceramic components and increases their costs, while the final results are not always satisfactory. For example, if during operation the materials suffer from considerable abrasion wear and thermocycling, it is quite difficult to protect them with coatings [7.36]. Thus, corrosion resistance of ceramics should be maintained by an optimization of their processing. There are three possible methods: reduction of open porosity and average pore sizes; certain amounts of necessary additives; lower contents of impurities. Basic information on the effect of porosity, additives and impurities on oxidation resistance is presented in Chap. 2. The investigation of silicon nitride materials was the most detailed one in this respect. Now we discuss the problem how additives can control the oxidation resistance of silicon nitride ceramics.

In addition to the major components, any raw material used for manufacturing the different products contains, calcium, magnesium, aluminium, strontium, nickel, cobalt, manganese, copper, titanium and other elements at a level of 1%–0.01% [7.37]. All these elements present in the material reduce the viscosity of silicate glass, and the larger the ionic radius of an element and the higher the oxidation rate, the lower the viscosity will be. Thus, the oxidation resistance of HS–130 (Norton Co., USA) (Fig. 2.17) will be higher than that of a similar HS–110 material which contains more impurities [7.38]. After oxidation impurities appear in the composition of a glassy phase or form different crystalline phases [7.39].

Thus, additives and impurities lowering the melting temperature of an oxide phase play a positive role by contributing to the formation of a continuous protective layer already at $\sim 1000°$ C. At the same time, at higher temperatures they abruptly increase the oxidation rate of the material. Therefore, for producing oxidation-resistant ceramics, one should clear raw materials from initial impurities and add precisely controlled amount of definite elements providing the required properties of the material.

Thus, the oxidation resistance of Si$_3$N$_4$ materials can be increased by adding zirconia, apart from magnesia [7.40]. A hot-pressed material with 1% MgO and 2% ZrO$_2$ displayed a much higher high-temperature oxidation resistance than the same material without zirconia. The authors of [7.40] believe that the addition of ZrO$_2$ retards oxygen diffusion through an oxide film and the transport of magnesium ions to the specimen surface. It has also been shown that the materials with ZrO$_2$ or ZrC additives possess a high oxidation resistance up to 1500° C. As described in [7.41,42], the extremely good performance of Si_3N_4–Al_2O_3–ZrO_2 composites in oxidizing environments is due to the following factors:

- absence of Zr–O–N phases which have been found to affect dramatically oxidation of sintered or hot-pressed silicon nitride materials containing ZrO_2;
- low amount of grain-boundary phase;
- decreased dissolution rate at the nitride/high viscous oxide film interface.

Yttria addition allows to get a crystalline grain-boundary phase [7.43], instead of a glassy one as in the case of MgO addition, and to increase the high-temperature strength and oxidation resistance of the materials. Care must be taken, however, to ensure that the phases formed by crystallization of the glass are not of the sort that provide a volume expansion upon oxidation, since this could lead to catastrophic cracking at intermediate temperatures where the plasticity of the oxide scale and the substrate are not sufficient to compensate for the volume change. A sudden release of nitrogen from apatite phases upon oxidation above 1000° C can also be a reason for cracking and explosive oxidation of Y-containing ceramics. Silicon-yttrium oxynitride formed due to interaction of Y_2O_3 with Si_3N_4 and SiO_2, which is always present on the surface of silicon nitride grains, were studied in [7.44]. It is recommended to maintain such processing conditions that the secondary phase consists of $Y_4Si_2O_7N_2$. Alumina and magnesia additions can facilitate this process. Such a material exhibits an extremely high oxidation resistance.

Materials with Y_2O_3 addition possess a high oxidation resistance at 1300°–1400° C [7.45]. In the range of 900°–1200° C the crystalline oxide film formed on the specimen surface does not have protective properties. This leads to intensive oxidation of the materials and to the deterioration of their mechanical properties [7.46]. The authors of [7.47] propose to call the temperature separating these two regions the critical transition temperature T_c. Above T_c the oxide layer is dense and has excellent protective properties. The oxidation kinetics obeys the parabolic law, the oxidation rate decreasing with the reduction of the additive content. Below T_c the surface layer contains connected pores and, thus, possesses poor protective properties. The oxidation kinetics is close to linear. For the materials of the system Si_3N_4–Y_2O_3 studied in [7.47], $T_c = 1200°$–1250° C. At temperatures below T_c a maximum oxidation rate is reached at 1000° C.

A dense oxide layer can be formed provided that the compounds decreasing the liquidus temperature of the grain-boundary phase are present in the composition of the material. One of them is Al_2O_3, The addition of 2% Al_2O_3 retards the crystallization of the grain-boundary phase, facilitates glass formation and reduces T_c by 200° C [7.47]. Such a material retains satisfactory strength only up to 1200° C. However, as opposed to materials with Y_2O_3 addition, it displays a good oxidation resistance at 1000° C, since this temperature is only slightly lower than its T_c. A hot-pressed material consisting of 93% Si_3N_4, 5% Y_2O_3 and 2% Al_2O_3 exhibited a good oxidation resistance during long-term tests (up to 30 days) at 1200° C [7.48]. There is information on the development of dense materials with 8% [7.49], 4% [7.50, 51] and 1% [7.52] Y_2O_3 which did not loose their properties at 1000°–1200° C. According to [7.47, 50, 51, 53], materials fabricated in the Si_3N_4–$Y_2Si_2O_7$–Si_2N_2O triangle display a high oxidation resistance. By changing the yttria content it is possible to control the oxidation behaviour of silicon oxynitride [7.54] and β-sialon [7.55] ceramics. Materials with beryllia addition [7.56] also have high oxidation resistance. A hot-pressed material of a $Si_{2.9}Be_{0.1}N_{3.8}O_{0.2}$ composition is reported to have a much higher oxidation resistance than commercially available Si_3N_4-based materials [7.56]. In this case, the activation energy of the oxidation process is almost twice as high as for MgO-doped silicon nitride materials (Table 2.5).

Depending on the material of milling bodies and the lining of the mill used for preparing the initial powder mixtures, they are contaminated with iron or tungsten carbide. Sometimes up to 2.5% of iron oxides are specially added to reaction-bonded materials to improve the process of nitridation [7.57]. After hot pressing or reaction bonding of the material these impurities are distributed along the grain-boundaries in their initial state, as a silicide or mixed iron-tungsten carbide [7.57]. Because of a high oxygen potential, they are oxidized and dissolved in the grain boundary phase decreasing its melting temperature. This results in a liquid phase and pits formed in the surface layer of the material at high temperatures and in the deterioration of high-temperature strength.

Analysing the data and information presented in Sect. 2.1, we may conclude that silicon nitride ceramics with the addition of elements decreasing the melting temperature and viscosity of a silicate phase (Mg, Al, B) exhibit higher oxidation resistance at intermediate temperatures, but these elements exert a negative effect on their high-temperature resistance. The addition of high-melting oxides (Y_2O_3, ZrO_2) increases high-temperature oxidation resistance but creates the danger of catastrophic oxidation in the range of intermediate temperatures. It is particularly dangerous when easily oxidizing oxynitride phases are formed along the grain boundaries. The oxidation protection and development of oxidation-resistant silicon-based ceramics are described in more detail, e.g., in [7.58]. We will not dwell here on other types of structural ceramics. During the examination of several classes of ceramics in Chap. 2, optimum compositions with the highest oxidation resistance were always mentioned. We should only note that the current level of knowledge does not allow to formulate any general principle of producing high oxidation-resistant ceramics based on different compounds. In every case we need a special approach accounting for the requirements to the material and its service conditions.

References

Chapter 1

1.1   D. Munz, T. Fett: *Mechanisches Verhalten keramischer Werkstoffe* (Springer, Berlin, Heidelberg 1989)

1.2   R.W. Davidge: *Mechanical Behaviour of Ceramics* (Cambridge University Press 1979)

1.3   M. Billy, J.G. Desmaison: High Temp. Technol. 4, 131-139 (1986)

1.4   Yu.G. Gogotsi, V.A. Lavrenko: Uspekhi Khimii 56, 1777-1797 (1987) [in Russian]

1.5   G.G. Gnesin: *Nonoxide Ceramic Materials* (Tekhnika, Kiev 1987) [in Russian]

1.6   S. Saito (ed.): *Fine Ceramics* (Elsevier, New York 1988)

1.7   Yu.G. Gogotsi: *Structural Ceramics: Manufacturing, Properties, Applications* (Znanie, Kiev 1990) [in Russian]

1.8   F. Thümmler: J. Europ. Cer. Soc. 6, 139-151 (1990)

1.9   F. Eisfeld (Hrsg.): *Keramik-Bauteile in Verbrennungsmotoren* (Vieweg, Braunschweig 1989)

1.10   E.T. Denisenko, T.V. Eremina, D.F. Kalinovich, L.I. Kuznetsova: Poroshk. Met. 3, 97-106 (1985) [in Russian]

1.11   G.V. Samsonov, O.P. Kulik, V.S. Polishchuk: *Processing and Methods of Analysis of Nitrides* (Naukova Dumka, Kiev 1978) [in Russian]

1.12   T.Ya. Kosolapova (ed.): *Nonmetallic Refractory Compounds* (Metallurgiya, Moscow 1985) [in Russian]

1.13   R.A. Andrievsky, I.I. Spivak: *Silicon Nitride and Materials on its Base* (Metallurgiya, Moscow 1984) [in Russian]

1.14   G.G. Gnesin: *Silicon Carbide Materials* (Metallurgiya, Moscow 1977) [in Russian]

1.15   D.S. Rutman, Yu.S. Toropov, S.Yu. Pliner, A.D. Neuimin, Yu.M. Polezhaev: *Refractory Zirconia Materials* (Metallurgiya, Moscow 1985) [in Russian]

1.16   Yu.L. Krasulin, V.N. Timofeev, S.M. Barinov: *Porous Structural Ceramics* (Metallurgiya, Moscow 1980) [in Russian]

1.17   U.R. Evans: *The Corrosion and Oxidation of Metals* (Edward Arnold, London 1960)

1.18   L.L. Shreir (ed.): *Corrosion* (Newnes-Butterworth, London, Boston 1977)

1.19   P. Kofstad: *High-Temperature Corrosion* (Elsevier, London 1988)

1.20   *Electrochemical Society Meeting. Corrosion of Ceramics and Advanced Materials:* J. Electrochem. Soc. 138, 148c-149c (1991)

1.21   R.E. Tressler, M. McNallan (eds.): "Corrosion and Corrosive Degradation of Ceramics" in *Ceramic Transactions*, Vol. 10 (Am. Cer. Soc., Westerville 1990)

1.22   R.J. Fordham (ed.): *High Temperature Corrosion of Technical Ceramics* (Elsevier, London 1990)

1.23   S. Somiya, M. Mitomo, M. Yoshimura (eds.): *Silicon Nitride-1* (Elsevier, London 1989)

1.24   D.A. Bonnell, T.Y. Tien (eds.): "Silicon Nitride: Preparation and Properties", in *Materials Science Forum*, Vol. 47 (Trans Tech Publications, 1989)

1.25   P.S. Kisly, M.A. Kuzenkova, N.I. Bodnaruk, B.L. Grabchuk: *Boron Carbide* (Naukova Dumka, Kiev 1988) [in Russian]

Chapter 2

2.1   G.R. Terwilliger: J. Am. Cer. Soc. **57**, 48-49 (1974)

2.2   K.G. Nickel, R. Danzer, G. Schneider, G. Petzow: *Korrosion und Oxidation von Hochleistungkeramik*, Symp. Materialforschung, Hamm/Westfalen, Sept. 12-19, 1988, pp. 611-630

2.3   M. Maeda, K. Nakamura, A. Tsuge: J. Mat. Sci. Lett. **8**, 195-197 (1989)

2.4   I.M. Kuleshov: Zh. Neorg. Khimii **4**, 488-491 (1959) [in Russian]

2.5   A.P. Pomytkin: "Kinetics of High-Temperature Oxidation of Titanium, Niobium, Chromium, Boron and Silicon Carbides in Oxygen"; Cand. Sci. Thesis, Institute for Problems of Materials Science (Kiev, 1976) [in Russian]

2.6   R.F. Vojtovich, E.A. Pugach: *Oxidation of Refractory Compounds*, Handbook (Metallurgiya, Moscow 1978) [in Russian]

2.7 · T.Ya. Kosolapova (ed.): *Manufacturing, Properties, and Applications of Refractory Compounds*. Handbook (Metallurgiya, Moscow 1986) [in Russian]

2.8   Ceramic Source '86-91, Vol. 1-6 (Am. Cer. Soc., Westerville 1985-1990)

2.9   R.F. Vojtovich: *Oxidation of Carbides and Nitrides* (Naukova Dumka, Kiev 1981) [in Russian]

2.10 V.A. Lavrenko, A.F. Alexeev, V.K. Kazakov, E.S. Lugovskaya: "High-Temperature Oxidation of Silicon Nitride" in *Refractory Nitrides* (Naukova Dumka, Kiev 1983) pp. 121-127 [in Russian]

2.11 S.S. Lin: J. Am. Cer. Soc. **58**, 160 (1975)

2.12 M. Wakamatsu, N. Takeuchi, T. Hattori, K. Watanabe, M. Ishikawa: "Oxidation of Sintered Si_3N_4 at 1400° C in Oxygen Atmosphere", in *Proc. UNITECR'89* (Am. Cer. Soc., Westerville 1989) pp. 1694-1703

2.13 M. Wakamatsu, N. Takeuchi, S. Shimuzu, T. Hattori, M. Oyama, H. Nanri, S. Ishida: "Oxidation of Sintered Si_3N_4 during Heating up to 1600° C under Oxygen Atmosphere", in *Proc. Materials Research Society Int. Symp.*, Vol. 4 (Mat. Res. Soc. 1989) pp. 283-288

2.14 A.G. Evans, R.W. Davidge: J. Mat. Sci. **5**, 314-325 (1970)

2.15 S.C. Singhal: J. Mat. Sci. **11**, 500-509 (1976)

2.16 B.-D. Kruse, G. Wilman, G. Hausner: Ber. DKG **53**, 349-351 (1976)

2.17 V.A. Lavrenko, Yu.G. Gogotsi: *Corrosion of Structural Ceramics* (Metallurgiya, Moscow 1989) [in Russian]

2.18 R.O. Williams, V.J. Tennery: J. Mat. Sci. **14**, 1567-1571 (1979)

2.19 B. Bergman, H. Heping: J. Europ. Cer. Soc. **6**, 3-8 (1990)

2.20 J.E. Sheehan: J. Am. Cer. Soc. **65**, 111-113 (1982)

2.21 I.Ya. Guzman, M.F, Lisov, Yu.N. Litvin: Trudy MKhTI Mendeleeva **118**, 59-67 (1981) [in Russian]

2.22 G.N. Babini, A. Bellosi, P. Vincenzini: La Ceramica **34**, No.3, 11-20 (1981)

2.23 G.N. Babini, A. Bellosi, P. Vincenzini: J. Mat. Sci. **18**, 231-244 (1983)

2.24 R.M. Horton: J. Am. Cer. Soc. **52**, 121-124 (1969)

2.25 N. Azuma, M. Maeda, K. Nakamura, M. Yamada: Yogyo-Kyokai-Shi **95**, 459-462 (1987) [in Japanese]

2.26 A. Giachello, P. Popper: Sci. Cer. **10**, 377-384 (1980)

2.27 V.A. Lavrenko, Yu. G. Gogotsi, O.D. Shcherbina: Sov. Powd. Met. Metal Cer. **24**, 710-713 (1985) [English transl.: Poroshk. Met. No.9, 62-66 (1985)]

2.28 A.F. Hampton, H.C. Graham: Oxid. Met. **10**, 239-253 (1976)

2.29 J. Echeberia, F. Castro: "Microstructure of the Oxide Layers Produced During Oxidation of Silicon Nitride", in *Euro-Ceramics*, ed. by G. de With, R.A. Terpstra, R. Metselaar, Vol. 3 (Elsevier, London 1989) pp. 527-532

2.30 S.C. Singhal: J. Am. Cer. Soc. **56**, 81-82 (1976)

2.31 N. Azuma, M. Maeda, K. Nakamura: Nagoya Kogyo Gijutsu Shikensho Hokoku **37**, 261-268 (1988) [in Japanese]

2.32 P. Andrews, F.L. Riley: J. Europ. Cer. Soc. **5**, 245-256 (1989)

2.33 R.A. Andrievsky, I.I. Spivak: *Silicon Nitride and Materials on its Base* (Metallurgiya, Moscow 1984) [in Russian]
2.34 W. Engel, F. Porz, F. Thümmler: Ber. DKG 52, 1296-1299 (1975)
2.35 T. Hirai, K. Niihara, T. Goto: J. Am. Cer. Soc. 63, 419-424 (1980)
2.36 G.A. Gogotsi: Some Results of Investigations of Mechanical Properties of Ceramics for Engine Parts (Inst. for Problems of Strength, Kiev 1983) [in Russian]
2.37 I.Ya. Guzman, M.F. Lisov: Zh. VKhO Mendeleeva 27, 553-557 (1982) [in Russian]
2.38 J.A. Palm, C.D. Greshkovich: Am. Cer. Soc. Bull. 59, 447-452 (1980)
2.39 S. Dutta, B. Buzek: J. Am. Cer. Soc. 67, 89-92 (1984)
2.40 F. Porz, F. Thümmler: J. Mat. Sci. 19, 1283-1295 (1984)
2.41 J.C. Uy: Am. Cer. Soc. Bull. 57, 735-737, 740 (1978)
2.42 M. Billy, P. Lortholary, M.-H. Negrier: Rev. int. hautes Temp. Refract. 15, 15-22 (1978)
2.43 V.A. Lavrenko, E.A. Pugach, A.B. Goncharuk, Yu.G. Gogotsi, G.V. Trunov: Sov. Powd. Met. Metal Cer. 23, 859-863 (1984) [English transl.: Poroshk. Met. No. 11, 50-54 (1984)]
2.44 U.R. Evans: *The Corrosion and Oxidation of Metals* (Edward Arnold, London 1960)
2.45 V.A. Lavrenko, Yu.G. Gogotsi, A.B. Goncharuk, A.F. Alexeev, O.N. Grigor'ev, O.D. Shcherbina: Sov. Powd. Met. Metal. Cer. 24, 207-210 (1985) [English transl.: Poroshk. Met. No. 3, 35-39 (1985)]
2.46 N.A. Toropov, V.P. Barzakovsky, V.V. Lapin, N.N. Kurtseva: *State Diagrams of Silicate Systems.* Handbook (Nauka, Leningrad 1969) [in Russian]
2.47 G.N. Babini, A. Bellosi, P. Vincenzini: J. Mat. Sci. 19, 1029-1042 (1984)
2.48 J.B. Veyret, M. Billy: "Oxidation of Hot–Pressed Silicon Nitride: Modelling", in *Euro-Ceramics*, ed. by G. de With, R.A. Terpstra, R. Metselaar, Vol. 3 (Elsevier, London 1989) pp. 512-516
2.49 P. Vincenzini, A. Bellosi, G.N. Babini: Cer. Int. 12, 133-145 (1986)
2.50 D.M. Mieskowski, W.A. Sanders: J. Am. Cer. Soc. 68, C160-C163 (1985)
2.51 J.T. Smith, C.L. Quackenbush: Am. Cer. Soc. Bull. 59, 529-532 (1980)
2.52 V.A. Lavrenko, Yu.G. Gogotsi, V.Zh. Shemet: Inorg. Mat. 21, 209-212 (1985) [English transl.: Izv. AN SSSR, Neorg. Mat. 21, 258-261 (1985)]
2.53 M.I. Mayer, F.L. Riley: J. Mat. Sci. 13, 1319-1328 (1978)
2.54 G.N. Babini, A. Bellosi, P. Vincenzini: Cer. Int. No. 3, 78 (1981)
2.55 A. Bellosi, P. Vincenzini, G.N. Babini: J. Mat. Sci. 23, 2348-2354 (1988)
2.56 Yu.G. Gogotsi, I.I. Osipova, I.I. Chugunova, V.Zh. Shemet: Sov. Powd. Met. Metal Cer. 26, 163-166 (1987) [English transl.: Poroshk. Met. No. 2, 75-79 (1987)]
2.57 M. Maeda, K. Nakamura, T. Ohkubo: Nippon Seramikkusu Kyokai Gakujutsu Ronbunshi 96, 1028-1032 (1988) [in Japanese]
2.58 V.A. Lavrenko, A.F. Alexeev: Cer. Int. 12, 25-31 (1986)
2.59 M. Pourshirazi: *Möglichkeiten und Grenzen keramischer Werkstoffe im Maschinenbau* (Technische Universität, Berlin 1987)
2.60 A. Giachello, P.C. Martinengo, G. Tommasini: Am. Cer. Soc. Bull. 59, 1212-1215 (1980)
2.61 J.C. Bressiani: "Untersuchung der Glasphasen Kristallisation in Y_2O_3-haltigen β-sialon Keramiken"; Dr. Naturwiss. Dissertation, Universität Stuttgart (1984)
2.62 M.H. Lewis, P. Barnard: J. Mat. Sci. 15, 443-448 (1980)
2.63 V.A. Gunchenko, V.N. Pavlikov, G.V. Trunov: Sov. Powd. Met. Metal Cer. 27, 470-474 (1988) [English transl.: Poroshk. Met. No. 6 (1988)]
2.64 S. Mason, M.H. Lewis, C.J. Reed: Cer. Eng. Sci. Proc. 10, 896 (1989)
2.65 J.G. Desmaison, F.L. Riley: J. Mat. Sci. 16, 2625-2628 (1981)
2.66 Yu.G. Gogotsi, L.V. Lavriv, O.D. Shcherbina: Inorg. Mat. 27, 2100-2102 (1991) [English transl.: Izv. AN SSSR, Neorg. Mat. No. 11, 2440-2452 (1991)]
2.67 R.K. Govila, J.A. Mangels, J.R. Baer: J. Am. Cer. Soc. 68, 413-418 (1985)
2.68 Y. Hasegawa, H. Tanaka, M. Tsutsumi, H. Suzuki: J. Cer. Soc. Jap. 88, 292-297 (1980)
2.69 B. Frisch, W.-R. Thile, R. Drumm: cfi/Ber. DKG 65, 277-284 (1988)
2.70 R.R. Miner: Ind. Heating 47, No.5, 14-15, 17 (1980)
2.71 G.G. Gnesin: *Silicon Carbide Materials* (Metallurgiya, Moscow 1977) [in Russian]

2.72 T.Ya. Kosolapova (ed.): *Nonmetallic Refractory Compounds* (Metallurgiya, Moscow 1985) [in Russian]

2.73 A. Yamaguchi: Taykabutsu **39**, 306-311 (1987) [in Japanese]

2.74 W.L. Vaughn, H. Maahs: J. Am. Cer. Soc. **73**, 1540-1543 (1990)

2.75 P. Schuster, E. Gugel: "Formation of Cristobalite from Silicon Carbide", in *Silicon Carbide*, ed. by G. Henish, R. Roy (Mir, Moscow 1972) pp. 301-309 [in Russian]

2.76 A.V. Dunikov, G.D. Semchenko, Yu.G. Gogotsi: Ogneupory No. 2, 14-17 (1984) [in Russian]

2.77 R.E. Tressler, J.A. Costello, Z. Zheng: "Oxidation of Silicon Carbide Ceramics", in *Proc. Symp. Ind. Heat Exchangers*, Pittsburg, November 6-8, 1985, pp. 307-314

2.78 S.C. Singhal: J. Mat. Sci. **11**, 1246-1253 (1976)

2.79 J. Schlichting: Ber. DKG **56**, 196-200 (1979)

2.80 I.N. Frantsevich (ed.): *Silicon Carbide, Properties and Fields of Application* (Naukova Dumka, Kiev 1975) [in Russian]

2.81 R. Pampuch, W. Ptak, S. Jonas, J. Stoch: "Formation of Ternary Si-O-C Phase(s) During Oxidation of SiC", in *Energy and Ceramics, Proc. 4th Int. Meet. on Modern Ceramic Technol.*, Saint-Vincent, May 28-31, 1979 (Elsevier, Amsterdam 1980) pp. 435-448

2.82 S.C. Singhal, F.F. Lange: J. Am. Cer. Soc. **58**, 433-435 (1975)

2.83 Yu. G. Gogotsi, A.V. Dunikov: Ogneupory No. 12, 19-22 (1986) [in Russian]

2.84 J. Schlichting, K. Schwetz: High Temp.- High Pres. **14**, 219-233 (1982)

2.85 V.A. Lavrenko, E.A. Pugach, S.I. Filipchenko, Yu.G. Gogotsi: Oxid. Met. **27**, 83-94 (1987)

2.86 Yu.G. Gogotsi: Izv. VUZov. Khimiya i Khim. Tekhnologiya **30**, No. 7, 54-57 (1987) [in Russian]

2.87 V.A. Lavrenko, E.A. Pugach, S.I. Filipchenko, Yu.G. Gogotsi: Sov. J. Superhard. Mat. 6, No. 3, 26-30 (1984) [English transl.: Sverkhtverdye Mat. No. 3, 21-24 (1984)]

2.88 M. Furukawa, T. Kitahira: Nippon Tungsten Rev. **14**, 30-43 (1981)

2.89 U. Ernstberger, H. Cohrt, F. Porz, F. Thümmler: cfi/ Ber. DKG **60**, 167-173 (1983)

2.90 O.P. Kulik, E.T. Denisenko, O.I. Krot: *High-Temperature Structural Ceramics. Manufacturing and Properties* (Inst. Problems of Mat. Sci., Kiev 1985)

2.91 K.F. Kuper, K.M. Georg, V. Hopkins: "Preparation and Oxidation of Aluminium Nitride", in *Special Ceramics* (Metallurgiya, Moscow 1968) pp. 39-63 [in Russian]

2.92 D. Suryanarayana: J. Am. Cer. Soc. **73**, 1108-1110 (1990)

2.93 V.A. Lavrenko, A.F. Alexeev, E.S. Lugovskaya, I.N. Frantsevich: Dokl. Akad. Nauk SSSR **255**, 641-645 (1980) [in Russian]

2.94 P. Boch, J.C. Glandus, J. Jarrige, J.P. Lecompte: Cer. Int. **8**, 34-40 (1982)

2.95 D. Launay, P. Goeuriot, F. Thevenot: Ann. Chim. **10**, 85-91 (1985)

2.96 A.V. Lysenko, E.A. Pugach, S.I. Filipchenko, G.G. Postolova, L.N. Lavrinenko, S.F. Korablev: Fizika i Tekhnologiya Vysokikh Davlenii **27**, 77-79 (1988) [in Russian]

2.97 C. Devin, J. Jarrige, J. Mexmain: Rev. Int. hautes Temp. Refract. **19**, 325-334 (1982)

2.98 Yu.G. Tkachenko, D.Z. Yurchenko, T.V. Dubovik: Poroshkovaya Met. No. 4, 88-90 (1985) [in Russian]

2.99 V.S. Bakunov, G.E. Val'yano, V.P. Vinogradov, A.V. Kirillin, A.V. Kostanovsky, E.P. Pakhomov: in *High-Temperature Nitrides and Materials on their Base* (Inst. Problems of Mat. Sci., Kiev 1985) pp. 72-77 [in Russian]

2.100 L.M. Jones, M.G. Nicholas: J. Mat. Sci. Lett. **8**, 265-266 (1989)

2.101 A.N. Pilyankevich, V.F. Britun, Yu.G. Tkachenko, V.K. Yulyugin: Poroshkovaya Met. No. 6, 76-81 (1984) [in Russian]

2.102 Yu.L. Krutsky, G.V. Galevsky, A.A. Kornilov: ibid. No.2, 47-50 (1983) [in Russian]

2.103 V.A. Lavrenko, Yu.G. Gogotsi, I.N. Frantsevich: Dokl. Akad. Nauk SSSR **275**, 114-117 (1984) [in Russian]

2.104 V.A. Lavrenko, Yu.G. Gogotsi: Oxid. Met. **29**, 193-202 (1988)

2.105 Yu.G. Gogotsi, V.V. Kovylyaev: Sov. Powd. Met. Metal Cer. **28**, 306-309 (1989) [English transl.: Poroshk. Met. No. 4, 71-75 (1989)]

2.106 L.N. Yefimenko, E.V. Lifshits, I.T. Ostapenko: Poroshkovaya Met. No. 4, 56-60 (1987) [in Russian]

2.107 V.A. Lavrenko, A.P. Pomytkin, P.S. Kisly, B.L. Grabchuk: Oxid. Met. **10**, 85-95 (1976)

2.108 R.C. Weast (ed.): *Handbook of Chemistry and Physics* (CRC Press, Boca Raton 1987)

2.109 Yu.G. Tkachenko, D.Z. Yurchenko, V.K. Yulyugin, V.N. Molyar, L.M. Murzin, E.S. Lugovskaya: Poroshkovaya Met. No. 12, 41-43 (1984) [in Russian]

2.110 A.V. Kurdyumov, A.N. Pilyankevich: *Phase Transformations in Carbon and Boron Nitride* (Naukova Dumka, Kiev 1979) [in Russian]

2.111 A.F. Alexeev, V.A. Lavrenko, V.S. Neshpor, I.N. Frantsevich: Dokl. Akad. Nauk SSSR **238**, 370-373 (1978) [in Russian]

2.112 R.G. Avarbe, B.N. Sharupin, E.V. Tupitsyna, R.Yu. Mamet'ev, I.P. Lesteva: in *High-Temperature Nitrides and Materials on their Base* (Inst. Problems of Mat. Sci., Kiev 1985) pp. 47-57 [in Russian]

2.113 V.A. Lavrenko, T.G. Protsenko, A.V. Bochko, A.F. Alexeev, E.S. Lugovskaya, I.N. Frantsevich: Dokl. Akad. Nauk SSSR **224**, 877-879 (1975) [in Russian]

2.114 A.V. Bochko, V.A. Lavrenko, V.L. Primachuk, T.G. Protsenko: Sverkhtverdye Mat. No. 1, 16-18 (1986) [in Russian]

2.115 R. Naslain, B. Harris (eds.): *Ceramic Matrix Composites* (Elsevier, London 1990)

2.116 Yu.G. Gogotsi, V.A. Lavrenko: High Temp. Technol. **6**, 79-87 (1988)

2.117 V.P. Yaroshenko, I.I. Osipova, Yu.G. Gogotsi, D.A. Pogorelova: "Einfluss der Dispersionshärtung durch TiN auf die Eigenschaften von Heissgepresstem Si_3N_4" in *Vorträge 9. Int. Pulvermet. Tagung*, Dresden, Okt. 23-25, 1989, B. 3 (Dresden, 1989) pp. 295-307

2.118 Yu.G. Gogotsi, V.K. Kazakov, V.A. Lavrenko, T.G. Protsenko, V.V. Shvaiko: Sov. J. Superhard. Mat. **10**, No. 1, 33-39 (1988) [English transl.: Sverkhtverdye Mat. No. 1, 27-33 (1988)]

2.119 Yu.G. Gogotsi, O.N. Grigor'ev, V.L. Tikush: Sov. Powd. Met. Met. Cer. **27**, 386-391 (1988) [English transl.: Poroshk. Met. No. 5, 60-66 (1988)]

2.120 Yu.G. Gogotsi, V.A. Lavrenko, T.G. Protsenko, A.I. Stegnij, V.L. Tikush: Sverkhtverdye Materialy No. 3, 22-26 (1989) [in Russian]

2.121 W. Wendlandt: *Thermal Methods of Analysis* (Wiley & Sons, New York 1974)

2.122 A. Bellosi, A. Tampieri, Yu-Zh. Liu: Mat. Sci. Eng. **A127**, 115–122 (1990)

2.123 M.A. Janney: Am. Cer. Soc. Bull. **66**, 322-324 (1987)

2.124 I. Ogawa, K. Kobayashi, S. Nishikawa: J. Mat. Sci. **23**, 1363-1367 (1988)

2.125 A.K. Tsapuk, L.G. Podobeda: Poroshkovaya Met. No. 3, 51-54 (1988) [in Russian]

2.126 K.L. Luthra, H.-D. Park: J. Am. Cer. Soc. **73**, 1014-1023 (1990)

2.127 P. Wang: "Kriechen und Oxidationsverhalten von SiC-whiskerverstärkten Al_2O_3/ZrO_2 Werkstoffen"; Dr. Naturwiss. Dissertation, Kernforschungszentrum Karlsruhe (1990)

2.128 K.L. Luthra: Cer. Eng. Sci. Proc. **8**, 649-653 (1987)

2.129 F. Lin, T. Marieb, A. Morrone, S. Nutt: in *High-Temperature / High-Performance Composites, Proc. Met. Res. Soc. Symp.*, Vol. 120 (Pittsburg 1988) pp. 323-332

2.130 M.P. Borom, M.K. Brun, L.E. Szala: Cer. Eng. Sci. Proc. **8**, 654-670 (1987)

2.131 C. Baudin, J.S. Moya: J. Am. Cer. Soc. **73**, 1417-1420 (1990)

2.132 Yu.G. Gogotsi, O.N. Grigor'ev, N.A. Orlovskaya, D.Yu. Ostrovoj, V.P. Yaroshenko: Ogneupory No. 11, 10-13 (1989) [in Russian]

2.133 H.Y. Liu, K.-L. Weisskopf, M.J. Hoffmann, G. Petzow: J. Europ. Cer. Soc. **5**, 123-133 (1989)

2.134 M. Backhaus-Ricoult: J. Am. Cer. Soc. **74**, 1793-1802 (1991)

Chapter 3

3.1 N.S. Jacobson, J.L. Smialek, D.S. Fox: "Molten Salt Corrosion of SiC and Si_3N_4", in *Handbook of Ceramics and Composites*, ed. by N.P. Cheremisinoff, Vol. 1 (Marcel Dekker, New York 1990) pp. 99-136

3.2   I.N. Frantsevich, R.F. Vojtovich, V.A. Lavrenko: *High-Temperature Oxidation of Metals and Alloys* (Gostekhizdat USSR, Kiev 1963) [in Russian]

3.3   J.R. Blachere, F.S. Petit: *High-Temperature Corrosion of Ceramics* (Noyes Corp., Park Ridge 1989)

3.4   S. Brooks, D.B. Meadowcroft: Proc. Brit. Cer. Soc. **26**, 237-250 (1978)

3.5   C.T. Sims, J.E. Palko: in *Proc. Workshop on Ceramics for Advanced Heat Engines* (Orlando 1977) pp. 287-294

3.6   R.I. Abraitis, D.F. Borshchevsky, K.V. Brinkene et al.: Trudy AN LitSSR, Ser. B, No. 1, 81-85 (1981) [in Russian]

3.7   R.I. Abraitis, K.V. Brinkene, V.K. Zabukas et al.: Trudy AN LitSSR, Ser. B, No. 6, 102-108 (1981) [in Russian]

3.8   Yu.G. Gogotsi, V.V. Shvajko, V.A. Lavrenko, N.N. Zudin, V.V. Kovylyaev, I.N. Frantsevich: Dokl. Akad. Nauk SSSR **286**, 901-903 (1986) [in Russian]

3.9   V.A. Lavrenko, Yu.G. Gogotsi: *Corrosion of Structural Ceramics* (Metallurgiya, Moscow 1989) [in Russian]

3.10   M.I. Mayer, F.L. Riley: J. Mat. Sci. **13**, 1319-1328 (1978)

3.11   N. Nassif: Thermochim. Acta **79**, 305-314 (1984)

3.12   M.G. Lawson, H.R. Kim, F.S. Pettit, J.R. Blachere: J. Am. Cer. Soc. **73**, 989-995 (1990)

3.13   G.A. Gogotsi, V.P. Zavada, Yu.G. Gogotsi: Cer. Int. **12**, 203-208 (1986)

3.14   D.S. Fox, N.S. Jacobson: J. Am. Cer. Soc. **71**, 128-138 (1988)

3.15   N.S. Jacobson, D.S. Fox: J. Am. Cer. Soc. **71**, 139-148 (1988)

3.16   D.S. Fox, J.L. Smialek: J. Am. Cer. Soc. **73**, 303-311 (1990)

3.17   N. Azuma, K. Nakamura: Chem. Express **3**, 755-758 (1988) [in Japanese]

3.18   N.S. Jacobson, C.A. Stearns, J.L. Smialek: Advanced Cer. Mat. **1**, 154-161 (1986)

3.19   N.S. Jacobson, J.R. Smialek: J. Am. Cer. Soc. **68**, 432-439 (1985)

3.20   N.J. Tighe, J. Sun, R.M. Hu: Cer. Eng. Sci. Proc. **8**, 805-811 (1987)

3.21   C.R. Brinkman, K. von Cook, B.E. Foster, R.L. Graves, W.K. Kahl, K.C. Liu, W.A. Simpson: Am. Cer. Soc. Bull. **68**, 1440-1445 (1989)

3.22   N.S. Jacobson: Oxid. Met. **31**, 91-103 (1989)

3.23   T.T. Lepisto, T.A. Mantyla: Cer. Eng. Sci. Proc. **10**, 658-667 (1989)

3.24   S.C. Singhal: J. Am. Cer. Soc. **56**, 81-82 (1976)

3.25   H.-E. Kim, A.J. Moorhead: J. Am. Cer. Soc. **73**, 3007-3014 (1990)

3.26   M. Maeda, K. Nakamura, T. Ohkubo: J. Mat. Sci. **23**, 3933-3938 (1988)

3.27   H.-E. Kim, D.W. Ready: "Active Oxidation of SiC in Low Dew-Point Hydrogen above 1400° C", in *Silicon Carbide '87*, ed. by J.C. Cawley, C.E. Semler, Ceramic Transactions, Vol. 2 (Am. Cer. Soc., Westerville 1989) pp. 301-312

3.28   J. Schlichting: Ber. DKG **56**, 196-200 (1979)

3.29   N.S. Jacobson, A.J. Eckel, A.K. Misra, D.L. Humphrey: J. Am. Cer. Soc. **73**, 2330-2332 (1990)

3.30   S. Vefsah, M. Billy, J. Jarrige: Ann. Chim. **10**, 79-83 (1985)

3.31   T. Sato, K. Haryu, T. Endo, M. Shimada: J. Mat. Sci. **22**, 2277-2280 (1987)

3.32   T. Sato, M. Shimada: J. Am. Cer. Soc. **68**, 356-359 (1985)

3.33   F.F. Lange, G.L. Dunlop, B.I. Davis: J. Am. Cer. Soc. **69**, 237-240 (1986)

3.34   T. Sato, S. Ohtaki, T. Endo et al.: Adv. Cer. **24A**, 501-508 (1988)

3.35   T.T. Lepistö, T.A. Mäntylä: Cer. Eng. Sci. Proc. **10**, 658-667 (1989)

3.36   M. Yoshimura, T. Noma, K. Kawabata, S. Somiya: J. Mat. Sci. Lett. **6**, 465-467 (1987)

3.37   M. Yoshimura: Am. Cer. Soc. Bull. **67**, 1950-1955 (1988)

3.38   W.C. Tripp, H.C. Graham: J. Am. Cer. Soc. **59**, 399-403 (1976)

3.39   T.Ya. Kosolapova (ed.): *Nonmetallic Refractory Compounds* (Metallurgiya, Moscow 1985) [in Russian]

3.40   T.E. Easler, R.B. Poeppel: in *High-Temperature Corrosion in Energy Systems*, ed. by M.F. Rothman (Metallurgical Soc. AIME, Warrendale 1984)

3.41   K.G. Nickel, P. Quirmbach, R. Brook, G. Petzow: Werkstoffe und Korrosion No. 12, 726-727 (1990)

3.42 J.E. Marra, E.R. Kreider, N.S. Jacobson, D.S. Fox: "The Behaviour of SiC and Si_3N_4 Ceramics in Mixed Oxidation/Chlorination Environments", in *Silicon Carbide '87*, ed. by J.C. Cawley, C.E. Semler, Ceramic Transactions, Vol. 2 (Am. Cer. Soc., Westerville 1989) pp. 275-287

3.43 J.E. Marra, E.R. Kreider, N.S. Jacobson, D.S. Fox: J. Am. Cer. Soc. **71**, 1067-1073 (1988)

3.44 D.-S. Park, M.J. McNallan, C. Park, W.W. Liang: J. Am. Cer. Soc. **73**, 1323-1329 (1990)

3.45 J.E. Marra, E.R. Kreider, N.S. Jacobson, D.S. Fox: J. Electrochem. Soc. **135**, 1571-1574 (1988)

3.46 S.Y. Ip, M.J. McNallan, M.E. Schreiner: "Oxidation of SiC Ceramics Heat Exchanger Materials in the Presence of Chlorine at 1300° C", in *Silicon Carbide '87*, ed. by J.C. Cawley, C.E. Semler, Ceramic Transactions, Vol. 2 (Am. Cer. Soc., Westerville 1989) pp. 289-299

3.47 B. Sommer, J. Sommer, H. Kainer: "Korrosion an SiC-Bauteilen bei Hochtemperaturbeanspruchung", in *Verschleiß und Korrosion*. Vorträge der Konferenz, Köln, 22.-23. Mai 1990 (DKG, 1990) pp. 207-227

3.48 L. Lay: *Corrosion Resistance of Technical Ceramics* (Her Majesty's Stationery Office, London 1983)

Chapter 4

4.1 G.V. Samsonov, O.P. Kulik, V.S. Polishchuk: *Processing and Methods of Analysis of Nitrides* (Naukova Dumka, Kiev 1978) [in Russian]

4.2 G.I. Postogvard, G.Yu. Parkhomenko, I.T. Ostapenko: Poroshkovaya Met. No.8, 28-30 (1983) [in Russian]

4.3 G.I. Postogvard, A.E. Makarenko, T.P. Ryzhova, I.T. Ostapenko: Poroshkovaya Met. No.1, 70-72 (1980) [in Russian]

4.4 V.A. Lavrenko, Yu.G. Gogotsi: Vestnik KPI. Khimicheskoe Mashinostroyeniye i Tekhnologiya 24, 22-24 (1987) [in Russian]

4.5 T. Sato, Y. Tokunaga, T. Endo, M. Shimida, K. Komeya, K. Nishida, M. Komatsa, T. Kameda: J. Mat. Sci. **23**, 3440-3446 (1988)

4.6 Yu.G. Gogotsi, V.N. Mikhatskaya, A.M. Pivak: Vestnik KPI. Khimicheskoe Mashinostroyeniye i Tekhnologiya 24, 26-30 (1987) [in Russian]

4.7 T.B. Tripp: J. Electrochem. Soc. **117**, 157-159 (1970)

4.8 G.V. Samsonov, I.M. Vinitsky: *Refractory Compounds*. Handbook (Metallurgiya, Moscow 1976)

4.9 T.Ya. Kosolapova (ed.): *Nonmetallic Refractory Compounds* (Metallurgiya, Moscow 1985) [in Russian]

4.10 M. Furukawa, T. Kitahira: Nippon Tungsten Rev. **14**, 30-43 (1981)

4.11 S.G. Seshadri, M. Srinivasan: J. Am. Cer. Soc. **71**, c72-c74 (1988)

4.12 R.W. Lashway, S.G. Seshadri, M. Shrinivasan: Lubr. Eng. **40**, 356-363 (1984)

4.13 A.M. Pivak, T.O. Malinovskaya, Yu.G. Gogotsi, L.F. Vasilenko: "Corrosion Resistance of Materials Based on Silicon and Boron Carbides in Electrolyte Solutions" in *New Technological Processes in Powder Metallurgy*, ed. by I.D. Radomyselsky (Inst. Problems of Mat. Sci., Kiev 1986) pp. 31-35 [in Russian]

4.14 H. Hirayama, T. Kawakubo, A. Goto, T. Kaneko: J. Am. Cer. Soc. **72**, 2049-2053 (1989)

4.15 A. Abid, R. Bensalem, B.J. Sealy: J. Mat. Sci. **21**, 1301-1304 (1986)

4.16 A.A. Pletyushkin, T.N. Sultanova, P.V. Volkova: Izv. AN SSSR, Neorg. Mat. **6**, 868-873 (1970) [in Russian]

4.17 W. Genthe, H. Hausner: cfi/Ber. DKG **67**, 6-11 (1990)

4.18 R. Morrell: *Handbook of Properties of Technical and Engineering Ceramics*. Part 1 (NPL, London 1989)

4.19   W. Genthe, H. Hausner: "Corrosion of Alumina in Acids" in *Euro-Ceramics*, ed. by G. de With, R.A. Terpstra, R. Metselaar, Vol. 3 (Elsevier, London 1989) pp. 463-467

4.20   R.E. Tressler, M.D. Meiser, J. Yonushonis: J. Am. Cer. Soc. **59**, 278-279 (1976)

4.21   S. Brooks, D.B. Meadowcroft: Proc. Brit. Cer. Soc. **26**, 237-250 (1978)

4.22   N.A. Pilipchuk, I.A. Kedrinsky, S.V. Morozov, T.V. Dubovik, G.V. Trunov, A.I. Itsenko: in *Refractory Nitrides* (Naukova Dumka, Kiev 1983) pp. 118-121 [in Russian]

4.23   Yu.G. Gogotsi, V.A. Lavrenko, I.N. Frantsevich: Dokl. Akad. Nauk SSSR **279**, 411-414 (1984) [in Russian]

4.24   T. Sato, Y. Kanno, T. Endo, M. Shimada: Yogyo-Kyokai-Shi **94**, 123-138 (1986) [in Japanese]

4.25   V.J. Jennings: "The Etching of Silicon Carbide" in Silicon Carbide, eds. G. Henish, R. Roy (Mir, Moscow 1972) pp. 279-289 [in Russian]

4.26   J.W. Cree, M.F. Amateau: Cer. Eng. Sci. Proc. **8**, 812-814 (1987)

4.27   T. Mitamura, E. Kogure, F. Noguchi et al.: Adv. Cer. **24A**, 109-118 (1988)

4.28   L. Lay: *Corrosion Resistance of Technical Ceramics* (Her Majesty's Stationery Office, London 1983)

Chapter 5

5.1   V.A. Lavrenko, Yu.G. Gogotsi: *Corrosion of Structural Ceramics* (Metallurgiya, Moscow 1989) [in Russian]

5.2   Ceramic Source '90, Vol. 5 (Am. Cer. Soc., Westerville 1989)

5.3   G. Orange, T. Epicier, G. Fantozzi: L'industrie Ceramique No. 12, 829-835 (1981)

5.4   Yu.G. Gogotsi, S.I. Sopenko, G.V. Trunov: Strength of Mat. No.1, 82-87 (1985) [English transl.: Problemy Prochnosti No.1, 69-72 (1985)]

5.5   R.K. Govila, J.A. Mangels, J.R. Baer: J. Am. Cer. Soc. **68**, 413-418 (1985)

5.6   F.F. Lange, B.I. Davis, A.G. Metcalf: J. Mater. Sci. **18**, 1497-1505 (1983)

5.7   V.A. Lavrenko, A.A. Chernovolenko, S.I. Sopenko, V.K. Zubov, A.F. Alexeev, Yu.G. Gogotsi, A.B. Goncharuk, V.V. Shvaiko: Strength of Mat. No.8, 1070-1074 (1986) [English Transl.: Problemy Prochnosti No.8, 67-70 (1986]

5.8   P. Goursat, A. Benyahia, J.L. Besson: Rev. Int. hautes Temp. Refract. **22**, 87-104 (1985)

5.9   P. Belair, J. Desmaison, Y. Bigay: "Influence of Heat Treatments under Oxidizing Atmospheres on the Mechanical Properties of RBSN Materials" in *Euro-Ceramics*, ed. by G. de With, R.A. Terpstra, R. Metselaar, Vol. 3 (Elsevier, London 1989) pp. 492-493

5.10   J.E. Ritter, P.A. Gennari, S.V. Nair, J.S. Haggerty, A. Lightfoot, Cer. Eng. Sci. Proc. **10**, 625-631 (1989)

5.11   J. Lamon, M. Boussuge: Sci. Ceram. **12**, 621-627 (1984)

5.12   D. Steinmann: Sprechsaal **121**, 909-920 (1988)

5.13   J.J. Mecholsky, P.F. Becher, S.W. Freiman: in *Surface Treatments for Improved Performance and Properties*, ed. by J.J. Burke, V. Weiss (Plenum Press, New York 1982)

5.14   F.F. Lange: J. Am. Cer. Soc. **63**, 38-40 (1980)

5.15   G. Das, M.G. Mendiratta, G.R. Cornish: J. Mater. Sci. **17**, 2486-2469 (1982)

5.16   T.E. Easler, R.C. Bradt, R.E. Tressler: J. Am. Cer. Soc. **65**, 317-320 (1982)

5.17   F.F. Lange, B.I. Davis, H.C. Graham: J. Am. Cer. Soc. **66**, C98-C99 (1983)

5.18   G. Wirth, W. Gebhard: Z. Werkstofftechn. **13**, 224-225 (1982)

5.19   D.R. Clarke: Mat. Sci. Forum **47**, 110-118 (1989)

5.20   H.-E. Kim, A.J. Moorhead: J. Am. Cer. Soc. **73**, 1443-1445 (1990)

5.21   H. Heigl, K. Heckel: cfi/Ber. DKG **60**, 69-73 (1983)

5.22   K.D. McHenry, R.E. Tressler: J. Mater. Sci **12**, 1272-1278 (1977)

5.23   P.F. Becher: J. Am. Cer. Soc. **66**, 120-121 (1983)

5.24   J.E. Siebels: in *Ceramics for High-Performance Application*, Proc. 6th Army Mater. Techn. Conf., Orcas Island, July 10-13, 1979 (New York 1983) pp. 793-804

5.25 F.F. Lange: J. Am. Cer. Soc. **53**, 290 (1970)
5.26 G.G. Trantina: Am. Cer. Soc. Bull. **57**, 440-443 (1978)
5.27 M.A. Walton, R.C. Bradt: Proc. Brit. Cer. Soc. **32**, 249-260 (1982)
5.28 K.D. McHenry, R.E. Tressler: Am Cer. Soc. Bull. **59**, 459-461 (1980)
5.29 S.G. Seshadri, M. Srinivazan: J. Am. Cer. Soc. **71**, c72-c74 (1988)
5.30 H.-E. Kim, A.J. Moorhead: J. Am. Cer. Soc. **73**, 1868-1872 (1990)
5.31 C. Devin, J. Jarrige, J. Mexmain: Rev. Int. hautes Temp. Refract. **19**, 325-334 (1982)
5.32 D. Launay, P. Goeuriot, F. Thevenot: Ann. Chim. **10**, 85-91 (1985)
5.33 D.J. Godfrey: Proc. Brit. Cer. Soc. **26**, 265-279 (1978)
5.34 W.C. Bourne, R.E. Tressler: Am. Cer. Soc. Bull. **59**, 443-452 (1980)
5.35 D.S. Fox, J.L. Smialek: J. Am. Cer. Soc. **73**, 303-311 (1990)
5.36 R. Förthmann, A. Naoumidis: "Einfluß der Gaskorrosion auf die Biegefestigkeit von Siliciumkarbid-Rohrwerkstoffen aus industrieller Fertigung", in *Verschleiß und Korrosion. Vorträge der Konferenz*, Köln, 22-23. Mai 1990 (DKG, 1990) pp. 282-290
5.37 G.A. Gogotsi, Yu.G. Gogotsi, V.P. Zavada, S.I. Sopenko: Strength of Mat. No.11, (1984) [English transl.: Problemy Prochnosti No.11, 17-21 (1984)]
5.38 G.A. Gogotsi, Yu.G. Gogotsi, A.V. Drozdov, O.D. Shcherbina: Sov. Powd. Met. Metal Cer. **25**, 746-750 (1986) [English transl.: Poroshk. Met. No. 9, 53-57 (1986)]
5.39 N.S. Jacobson, J.L. Smialek, D.S. Fox: "Molten Salt Corrosion of SiC and Si_3N_4", in *Handbook of Ceramics and Composites*, ed. by N.P. Cheremisinoff, Vol. 1 (Marcel Dekker, New York 1990) pp. 99-136
5.40 J.L. Smialek, N.S. Jacobson: J. Am. Cer. Soc. **69**, 741-752 (1986)
5.41 T. Sato, Y. Koike, T. Endo, M. Shimada: J. Mater. Sci. **23**, 1405-1410 (1988)
5.42 J.J. Swab, G.L. Leatherman: J. Europ. Cer. Soc. **5**, 333-340 (1989)
5.43 C.R. Brinkman, K. von Cook, B.E. Foster, R.L. Graves, W.K. Kahl, K.C. Liu, W.A. Simpson: Am. Cer. Soc. Bull. **68**, 1440-1445 (1989)
5.44 P.F. Becher: J. Mater. Sci. **19**, 2805 (1984)
5.45 M.K. Ferber, J. Ogle, V.J. Tennery, T. Henson: J. Am. Cer. Soc. **68**, 191 (1985)
5.46 R.E. Tressler, M. McNallan (eds.): *Corrosion and Corrosive Degradation of Ceramics*, in Ceramic Transactions, Vol. 10 (Am. Cer. Soc., Westerville 1990)
5.47 J.J. Petrovic, L.A. Jacobson, P.K. Talty, A.K. Vasudevan: J. Am. Cer. Soc. **58**, 113-116 (1975)
5.48 G.A. Gogotsi, V.P. Zavada, Yu.G. Gogotsi: Ceram. Int. **12**, 203-208 (1986)
5.49 J.W. Cree, M.F. Amateau: Cer. Eng. Sci. Proc. **8**, 812-814 (1987)
5.50 J.R. McLaren, G. Tappin, R.W. Davidge: Proc. Brit. Cer. Soc. **20**, 259-274 (1972)

Chapter 6

6.1 R.W. Davidge, A.G. Evans: Mater. Sci. Eng. **6**, 281-298 (1970)
6.2 V.I. Trefilov, Yu.V. Milman, I.V. Gridneva: Crystal Res. Technol. **19**, 413-421 (1984)
6.3 F. Thevenot: "Boron Carbide – a Comprehensive Review" in *Euro-Ceramics*, ed. by G. de With, R.A. Terpstra, R. Metselaar, Vol. 2 (Elsevier, London 1989) pp. 2.1-2.25
6.4 G. Orange, T. Epicier, G. Fantozzi: L'industrie Ceramique No. 12, 829-835 (1981)
6.5 R.K. Govila: Am. Cer. Soc. Bull. **65**, 1287-1292 (1986)
6.6 Yu.G. Gogotsi, O.N. Grigorjew, W.P. Yaroschenko: Silikattechnik **41**, 156-160 (1990)
6.7 G. Grathwohl, F. Thümmler: Ceramurgia Int. **6**, No 2, 43-50 (1980)
6.8 U. Ernstberger, G. Grathwohl, F. Thümmler: Int. J. High Tech. Cer. **3**, 43-61 (1987)
6.9 O. Van Der Biest, S. Valkiers, L. Garguet, T. Tambuyser, I. Baele: Brit. Cer. Proc. **39**, 33-44 (1987)
6.10 G. Grathwohl, F. Thümmler: Ber. DKG **52**, 268-270 (1975)
6.11 J.M. Birch, B. Wilshire: J. Mater. Sci. **13**, 2627-2636 (1978)
6.12 D. Jiang, G. Kuang, Z. Pan, S. Tan, J. Mao, S. Ke: "Oxidation Behaviour and High-Temperature Strength of Hot-Pressed SiC with Various Sintering Aids", in *High-Tech*

Ceramics, ed. by P. Vincenzini, Materials Science Monographs, 38C (Elsevier, Amsterdam 1987)

6.13 G.A. Gogotsi, Yu.G. Gogotsi, D.Yu. Ostrovoj, O.V. Ivashchenko: Ogneupory No. 2, 19-24 (1989) [in Russian]

6.14 G.A. Gogotsi, Yu.G. Gogotsi, D.Yu. Ostrovoj: Ogneupory No. 10, 27-30 (1989) [in Russian]

6.15 R.W. Davidge: *Mechanical Behaviour of Ceramics* (Cambridge University Press 1979)

6.16 Yu.G. Gogotsi, V.A. Lavrenko: Ogneupory No. 5, 25-28 (1985) [in Russian]

6.17 U. Ernstberger, H. Cohrt, F. Porz, F. Thümmler: cfi/ Ber. DKG 60, 167-173 (1983)

6.18 G.G. Trantina: Am. Cer.Soc. Bull. 57, 440-443 (1978)

6.19 F.F. Lange: J. Am. Cer. Soc. 53, 290 (1970)

6.20 F. Thümmler, G. Grathwohl: "High-Temperature Oxidation and Creep of Si_3N_4- and SiC-Based Ceramics and Their Mutual Interaction" in *Proc. MRS Int. Meeting on Advanced Materials* (Mat. Res. Soc. 1989) pp. 237-253

6.21 G.A. Gogotsi, S.A. Firstov, A.D. Vasil'ev, Yu.G. Gogotsi, V.V. Kovylyaev: Sov. Powd. Met. Metal Cer. 26, 589-594 (1987) [English transl.: Poroshk. Met. No. 7, 84-90 (1987)]

6.22 G.A. Gogotsi, Ya.L. Groushevsky, O.B. Dashevskaya, Yu.G. Gogotsi, V.A. Lavrenko: J. Less.-Com. Met. 117, 225-230 (1986)

6.23 G.A. Gogotsi, Yu.G. Gogotsi, D.Yu. Ostrovoj: J. Mat. Sci. Lett. 7, 814-816 (1988)

6.24 G.A. Gogotsi, Yu.G. Gogotsi, V.V. Kovylyaev, D.Yu. Ostrovoj, V.Ya. Ivas'kevich: Sov. Powd. Met. Metal Cer. 28, 487-490 (1987) [English transl.: Poroshk. Met. No 6, 77-82 (1989)]

6.25 G. De With: J. Mater. Sci. 19, 457-466 (1984)

6.26 G.A. Gogotsi, Yu.G. Gogotsi, V.P. Zavada, V.V. Traskovsky: Strength of Mat. 21, 918-922 (1989) [English transl.: Problemy Prochnosti No. 7, 76-80 (1989)]

6.27 N.S. Jacobson, J.L. Smialek, D.S. Fox: "Molten Salt Corrosion of SiC and Si_3N_4", in *Handbook of Ceramics and Composites,* ed. by N.P. Cheremisinoff, Vol. 1 (Marcel Dekker, New York 1990) pp. 99-136

6.28 Yu.G. Gogotsi: Fiz.-Khim. Mekhanika Mat. 23, No 4, 73-82 (1987) [in Russian]

6.29 O.V. Mazurin, M.V. Streltsova, T.P. Shvaiko-Shvaikovskaya: *Properties of Glasses and Glass Melts,* Vol. 1 (Nauka, Leningrad 1973) [in Russian]

6.30 Yu.G. Gogotsi, V.P. Zavada, N.N. Zudin, V.V. Ivzhenko, V.V. Traskovsky: Sverkhtv. Mat. No 3, 25-29 (1990) [in Russian]

6.31 L.N. Petrov: *Stress Corrosion* (Vishcha Shkola, Kiev 1986) [in Russian]

6.32 D.F. Dailly, G.W. Hastings, S. Lash: Proc. Brit. Cer. Soc. 31, 191-200 (1981)

6.33 M.E. Gulden, A.G. Metcalf: J. Am. Cer. Soc. 59, 391-396 (1976)

6.34 A.G. Metcalf: in *Proc. Workshop on Ceramics for Advanced Heat Engines,* Orlando, Jan. 24-26, 1977, pp. 279–286

6.35 J.E. Ritter, S.M. Wiederhorn, N.J. Tighe, E.R. Fuller: in *Ceramics for High-Performance Application,* Proc. 6th Army Mater. Techn. Conf., Orcas Island, July 10-13, 1979 (New York, 1983) pp. 503-533

6.36 R.J. Charles, W.B. Hillig: in *High-Strength Materials,* ed. by V.F. Zackey (Wiley, New York 1965) pp. 682-705

6.37 S.M. Wiederhorn, E.R. Fuller, R. Thomson: Metal Sci. No. 8/9, 450-458 (1980)

6.38 R. Thomson: J. Mater. Sci. 15, 1014-1026 (1980)

6.39 E.R. Fuller, R. Thomson: J. Mater. Sci. 15, 1027-1034 (1980)

6.40 A.S. Krausz, J. Mshana, K. Krausz: Eng. Fract. Mech. 13, 759-766 (1980)

6.41 M.G. Gee, L.N. McCartney: Proc. Brit. Cer. Soc. 32, 133-147 (1982)

6.42 G.P. Cherepanov: *Mechanics of Brittle Fracture* (Nauka, Moscow 1974) [in Russian]

6.43 R. Adams, P.W. McMillan: J. Mater. Sci. 12, 643-657 (1977)

6.44 C. Rinaldi: Ceramurgia 8, 20-24 (1978)

6.45 T.E. Easler, R.C. Bradt, R.E. Tressler: J. Am. Cer. Soc. 65, 317-320 (1982)

6.46 G.D. Quinn: Cer. Eng. Sci. Proc. 3, 77-98 (1982)

6.47 R.E. Tressler, M. McNallan (eds.): *Corrosion and Corrosive Degradation of Ceramics,* in Ceramic Transactions, Vol. 10 (Am. Cer. Soc., Westerville 1990)

6.48 G. Ziegler: Proc. Brit. Cer. Soc. **32**, 213-225 (1982)

6.49 G.R. Terwilliger: J. Am. Cer. Soc. **57**, 48-49 (1974)

6.50 S.A. Bortz: SAMPE J. **1**, 16-31 (1981)

6.51 W.C. Bourne, R.E. Tressler: Am. Cer. Soc. Bull. **59**, 443-452 (1980)

6.52 P. Belair, J. Desmaison, Y. Bigay: "Influence of Heat Treatments under Oxidizing Atmospheres on the Mechanical Properties of RBSN Materials", in *Euro-Ceramics*, ed. by G. de With, R.A. Terpstra, R. Metselaar, Vol. 3 (Elsevier, London 1989) pp. 492-493

6.53 Yu.G. Gogotsi, V.P. Zavada, V.V. Traskovsky: Strength of Mat. No. 11, 1555-1559 (1987) [English transl.: Problemy Prochnosti No. 11, 82-86 (1987)]

6.54 Yu.G. Gogotsi, V.P. Zavada, V.V. Traskovsky: Fiz.-Khim. Mekhanika Mat. **24**, No 6, 17-21 (1988) [in Russian]

6.55 G.A. Gogotsi, Yu.G. Gogotsi, V.P. Zavada, V.V. Traskovsky: Cer. Int. **15**, 305-310 (1989)

6.56 G. Himsolt: Z. Werkstofftechnik **13**, 225 (1982)

6.57 G.S. Pisarenko, Yu.G. Gogotsi, V.P. Zavada, V.V. Traskovsky: Dokl. Akad. Nauk SSSR **304**, 1184-1187 (1989) [in Russian]

6.58 Yu.G. Gogotsi: "Effect of Oxidation on Fracture of Nonoxide Ceramics", in *Fracture Mechanics and Physics of Brittle Materials* (IPM AN UkrSSR, Kiev 1990) pp. 115-122 [in Russian]

6.59 A.L. Rabinovich: *Introduction to Mechanics of Reinforced Polymers* (Nauka, Moscow 1970)

6.60 R.K. Govila, J.A. Mangels, J.R. Baer: J. Am. Cer. Soc. **68**, 413-418 (1985)

6.61 D. Steinmann: Sprechsaal **122**, 215-223 (1989)

6.62 S.H. Knickerbocker, A. Zangvill, S.D. Brown: J. Am. Cer. Soc. **67**, 365-369 (1984)

6.63 G. Ziegler: "Thermo-Mechanical Properties of Silicon Nitride and Their Dependence on Microstructure", in *Preparation and Properties of Silicon Nitride-Based Materials*, ed. by D.A. Bonnell, T.Y. Tien, Materials Science Forum, Vol. 47 (Trans. Tech. Publications, 1989) pp. 162-203

6.64 R. Kossowsky: J. Am. Cer. Soc. **56**, 531-535 (1973)

6.65 M. Boussuge, D. Broussaud, J. Lamon: Proc. Brit. Cer. Soc. **32**, 205-212 (1982)

6.66 J.L. Henshall, D.J. Rowcliff, J.W. Edington: J. Amer. Cer. Soc. **62**, 36-41 (1979)

6.67 K.D. McHenry, R.E. Tressler: J. Mater. Sci **12**, 1272-1278 (1977)

6.68 J.L. Henshall: Proc. 5th Int. Conf. on Fracture, Cannes, March 29 - April 3, 1981 (Cannes, 1982) pp. 1541-1549

6.69 K.D. McHenry, R.E. Tressler: J. Am. Cer. Soc. **63**, 152-156 (1980)

6.70 G.A. Gogotsi, Yu.G. Gogotsi, D.Yu. Ostrovoj: Ogneupory No. 10, 27-30 (1989) [in Russian]

6.71 M.A. Walton, R.C. Bradt: Proc. Brit. Cer. Soc. **32**, 249-260 (1982)

6.72 D.E. Schwab, D.M. Kotchick: Am. Cer. Soc. Bull. **59**, 805-808, 813 (1980)

6.73 D. Avigdor, S.D. Brown: J. Am. Cer. Soc. **61**, 97-99 (1978)

6.74 N.L. Hecht, S.D. Jang, D.E. McCullum: Adv. Ceram. **24A**, 133-144 (1988)

6.75 R.H. Dauskardt, R.O. Ritchie: Cer. Eng. Sci. Proc. **10**, 1146 (1989)

6.76 M.V. Swain, V. Zelizko: Adv. Ceram. **24B**, 595-606 (1988)

6.77 F.F. Lange: J. Am. Cer. Soc. **69**, 237-240 (1986)

6.78 H. Olapinski, U. Dworak, W. Burger: "High-Temperature Durability of Zirconia", in *Ceramic Materials and Components for Engines*, Proc. 2nd Int. Symp., Lübeck-Travemünde, April 14-17, 1986, ed. by W. Bunk, H. Hausner (Verlag DKG, 1986) pp. 535-549

6.79 G.A. Gogotsi, Yu.G. Gogotsi, V.P. Zavada, V.V. Traskovsky: Mat. High Temp. **9**, 209-216 (1991)

6.80 M.A. Lamkin, F.L. Riley, R.J. Fordham: "Hot Corrosion of Silicon Nitride in the Presence of Sodium and Vanadium Compounds", in *High-Temperature Corrosion of Technical Ceramics*, ed. by R.J. Fordham (Elsevier, London 1990) pp. 181-191

6.81 Technology Update. Am. Cer. Soc. Bull. **67**, 1886 (1988)

6.82   J.J. Swab, G.L. Leatherman: J. Europ. Cer. Soc. 5, 333-340 (1989)
6.83   Yu. Gogotsi, D. Ostrovoj, V. Traskovsky: "Deformation and Creep of Silicon Nitride–
       Matrix Composites", in *Mechanics of Creep Brittle Materials – 2*, ed. by A.C.F. Cocks,
       A.R.S. Ponter (Elsevier, London 1991) pp. 230-241
6.84   C. Razim, C. Kaniut, K.-H. Thiemann: "Keramik als Werkstoff für verschleiss-
       beanspruchte Bauteile", in *Verschleiß und Korrosion*. Vorträge der Konferenz, Köln,
       22-23. Mai 1990 (DKG, 1990) pp. 1-19
6.85   S.A. Horton, J. Denape, D. Broussaud, D. Dowson, F.L. Riley, N. Wallbridge: "The
       Wear Behaviour of Sialon and Silicon Carbide Ceramics in Sliding Contact", in *Nonox-
       ide Technical and Engineering Ceramics*, ed. by S. Hampshire (Elsevier, London 1986)
       pp. 281-298
6.86   A. Bellosi: Adv. Cer. Glass 1, 18-23 (1990)
6.87   H. Kolaska, K. Dreyer: "Verschleissverhalten von Schneidkeramik", in *Verschleiß und
       Korrosion. Vorträge der Konferenz*, Köln, 22-23. Mai 1990 (DKG, 1990) SS. 111-131
6.88   Yu.G. Gogotsi, V.V. Koval'chuk, V.V. Zametailo, R.G. Timchenko, I.A. Kossko, A.M.
       Koval'chenko: Trenie i Iznos 10, 295-301 (1989) [in Russian]
6.89   Yu.G. Gogotsi, A.M. Koval'chenko, I.A. Kossko, V.P. Yaroshenko: Trenie i Iznos 11,
       661-667 (1990) [in Russian]
6.90   A.M. Koval'chenko, I.A. Kossko, Yu.G. Gogotsi: "Investigation of Nonoxide Ceramics-
       Steel Interaction in Friction" in *Physical Materials Science and Physico-Chemical
       Fundamentals for Development of New Materials*, ed. by V.V. Skorokhod (IPM AN
       UkrSSR, Kiev 1989) pp. 108-113 [in Russian]
6.91   Yu.G. Gogotsi, O.N. Grigor'ev, A.M. Koval'chenko, I.A. Kossko, V.P. Yaroshenko: TIZ
       Int. Powd. Mag. 113, 701-702 (1989)
6.92   A.N. Pilyankevich, V.F. Britun, Yu.G. Tkachenko, V.K. Yulyugin: Poroshk. Met. No.
       8, 93-97 (1986) [in Russian]
6.93   Yu.G. Gogotsi, A.M. Koval'chenko, I.A. Kossko, V.P. Yaroshenko: "High-Speed Sliding
       of Ceramics Against Steel" in *Urgent Problems of Materials Science*, ed. by V.V.
       Skorokhod (IPM AN UkrSSR, Kiev 1991) pp. 101-106 [in Russian]
6.94   A.G. Evans, T.G. Langdon: "Structural Ceramics" in *Progress in Materials Science*,
       Vol. 21 (Pergamon Press, Oxford 1976) pp. 171-441
6.95   G.G. Trantina: J. Am. Cer. Soc. 62, 377-380 (1979)
6.96   A.G. Evans: Int. J. Fracture 9, 267-275 (1973)
6.97   A.G. Evans, L.R. Russell, D.W. Richerson: Metall. Trans. A. 6, 707-716 (1975)
6.98   D.P.H. Hasselman, E.P. Chen, C.L. Ammann et al.: J. Am. Cer. Soc. 58, 513-516 (1975)
6.99   J.C. Uy: Proc. Workshop on Ceramics for Advanced Heat Engines, Orlando, Jan. 24-
       26, 1977, pp. 259-267
6.100 A. Paluszny: Proc. Workshop on Ceramics for Advanced Heat Engines, Orlando, Jan.
       24-26, 1977, pp. 231-239
6.101 G.A. Gogotsi, V.P. Zavada: Problemy Prochnosti No. 2, 53-56 (1985) [in Russian]
6.102 V.P. Zavada, S.G. Nikolsky, V.A. Strizhalo, V.P. Terent'ev: Problemy Prochnosti No.
       8, 41-46 (1988) [in Russian]
6.103 R.K. Govila: in *Ceramics for High-Performance Application*, Proc. 6th Army Mater.
       Techn. Conf., Orcas Island, July 10-13, 1979 (New York, 1983) pp. 535-567
6.104 R.C. Bradt, F.F. Lange, D.P.H. Hasselman, A.G. Evans (eds.): *Fracture Mechanics of
       Ceramics*, Vols. 1-6 (Plenum Press, New York 1974-1982)

Chapter 7

7.1   A.V. Emyashev: *Gas-Phase Metallurgy of Refractory Compounds* (Metallurgiya, Mos-
      cow 1987) [in Russian]
7.2   Yu.S. Borisov, A.L. Borisova: *Plasma-Sprayed Powder Coatings* (Tekhnika, Kiev 1986)
      [in Russian]

7.3   R.A. Andrievsky, I.I. Spivak: *Silicon Nitride and Materials on its Base* (Metallurgiya, Moscow 1984) [in Russian]

7.4   A.V. Rzhanov (ed.): *Silicon Nitride in Electronics*, in Materials Science Monographs, Vol. 34 (Amsterdam, Elsevier 1987)

7.5   D.P. Stinton, T.M. Besmanm, R.A. Lowden: Am. Cer. Soc. Bull. **67**, 350-355 (1988)

7.6   J.J. Stiglich, D.G. Bhat, R.A. Holzl: Ceram. Int. **6**, 3-10 (1980)

7.7   T. Hirai, K. Niihara, T. Goto: J. Am. Cer. Soc. **63**, 419-424 (1980)

7.8   T. Hirai: in *Emergent Process Meth. High-Technol. Ceram.* Proc. Conf., Raleigh, Nov. 8-10, 1982 (New York, 1984) pp. 329-345

7.9   J.M. Blocher, M.F. Browning, D.M. Barrett: in *Emergent Process Meth. High-Technol. Ceram.* Proc. Conf., Raleigh, Nov. 8-10, 1982 (New York, 1984) pp. 299-316

7.10   R.C. Hendricks, G. McDonald, R.L. Mullen: Cer. Eng. Sci. Proc. **4**, 802-809 (1983)

7.11   G.G. Gnesin: *Silicon Carbide Materials* (Metallurgiya, Moscow 1977) [in Russian]

7.12   D.M. Kotchick, C.J. Dobos: in *High-Temperature Protective Coatings*, Proc. 112th AIME Annual Meet., Atlanta, March 7-8, 1983 (Warrendale 1983) pp. 281-292

7.13   G. Grathwohl, F. Porz, F. Thümmler: Ber. DKG **53**, 346-348 (1976)

7.14   G. Grathwohl, F. Porz, F. Thümmler: Radex-Rdsch. No. 2, 105-109 (1977)

7.15   T. Narushima, T. Goto, Y. Iguchi, T. Hirai: J. Am. Cer. Soc. **73**, 3580-3584 (1990)

7.16   K.T. Scott: in *Material Process in Industry* (London, 1982) pp. 99-105

7.17   V.N. Lyasnikov, V.S. Ukrainsky, G.F. Bogatyrev: *Plasma Spraing of Coatings for Manufacturing of Electronic Devices* (Saratov University, Saratov 1985) [in Russian]

7.18   G.I. Zhuravlev: *Chemistry and Technology of Thermally Resistant Inorganic Coatings* (Khimiya, Leningrad 1975) [in Russian]

7.19   P. Chagnon, P. Fauchis: Cer. Int. **10**, 119-131 (1984)

7.20   J.R. Spann, R.W. Rice, W.S. Coblenz, W.J. McDonough: in *Emergent Process Meth. High-Technol. Ceram.* Proc. Conf., Raleigh, Nov. 8-10, 1982 (New York, 1984) pp. 473-503

7.21   N.R. Shankar, C.C. Berndt, H. Herbert: Cer. Eng. Sci. Proc. **4**, 784-791 (1983)

7.22   S.C. Singhal (ed.): *High-Temperature Protective Coatings*, Proc. 112th AIME Annual Meet., Atlanta, March 7-8, 1983 (Warrendale, 1983)

7.23   O.J. Gregory, M.H. Richman: J. Am. Cer. Soc. **67**, 335-340 (1984)

7.24   J.R. Price, R.E. Gildersleeve, M. Van Roode, C.E. Smeltzer: in *Corrosion of Ceramic Materials Workshop,* ed. by B.K. Kennedy (Pennsylvania State University, University Park, 1987) pp. 81-89

7.25   P. Hancock: "Degradation Processes for Ceramic Coatings", in *Ceramic Coatings for Heat Engines*, ed. by I. Kvernes, W.J.G. Bunk, J.G. Wurm, Proc. Conf., Strasbourg, France, Nov. 26-28, 1985 (MRS, 1985) pp. 163-179

7.26   Y. Ynomata: J. Cer. Soc. Jap. **83**, No 3, 9-11 (1975)

7.27   V.A. Lavrenko, Yu.G. Gogotsi: *Corrosion of Structural Ceramics* (Metallurgiya, Moscow 1989) [in Russian]

7.28   W.S. Coblenz, G.H. Wiseman, P.B. Davis, R.W. Rice: in *Emergent Process Meth. High-Technol. Ceram.* Proc. Conf., Raleigh, Nov. 8-10, 1982 (New York, 1984) pp. 271-285

7.29   D. Seyferth, G.H. Wiseman, C. Prud'homme: in *Emergent Process Meth. High-Technol. Ceram.* Proc. Conf., Raleigh, Nov. 8-10, 1982 (New York, 1984) pp. 263-269

7.30   N. Nassif: Thermochim. Acta **79**, 305-314 (1984)

7.31   V.A. Lavrenko, E.A. Pugach, A.B. Goncharuk, Yu.G. Gogotsi, G.V. Trunov: Sov. Powd. Met. Metal Cer. **23** (1984) [English transl.: Poroshk. Met. No. 11, 50-54 (1984)]

7.32   F.F. Lange, B.I. Davis, A.G. Metcalf: J. Mater. Sci. **18**, 1497-1505 (1983)

7.33   Yu.G. Gogotsi, A.G. Gogotsi, O.D. Shcherbina: Sov. Powd. Met. Metal Cer. **25**, 388-391 (1986) [English transl.: Poroshk. Met. No. 5, 40-44 (1986)]

7.34   D.R. Clarke: J. Am. Cer. Soc. **67**, 455-459 (1984)

7.35   C.J. McHargue, C.S. Yust: J. Am. Cer. Soc. **67**, 117-123 (1984)

7.36   Yu.G. Gogotsi: Poverkhnost No. 6, 143-149 (1989) [in Russian]

7.37   L.G. Podobeda: Poroshk. Met. No. 1, 75-80 (1979) [in Russian]

7.38 R. Kossowsky: J. Am. Cer. Soc. **56**, 531-535 (1973)

7.39 A.J. Kiehle, L.K. Heung, P.J. Gielisse, T.J. Rockett: J. Am. Cer. Soc. **58**, 17-20 (1975)

7.40 A.F. Hampton, H.C. Graham: Oxid. Met. **10**, 239-253 (1976)

7.41 A. Bellosi, P. Vincenzini, G.N. Babini: J. Mat. Sci. **23**, 2348-2354 (1988)

7.42 L.K.L. Falk: "Microstructure and Oxidation Behaviour of Si_3N_4/ZrO_2 Ceramics", in *Structural Ceramics. Processing, Microstructure and Properties*, ed. by J.J. Bentzen et al. (Riso Nat. Lab., Roskilde 1990) pp. 277-282

7.43 A. Tsuge, K. Nishida, M. Komatsu: J. Am. Cer. Soc. **58**, 323-326 (1975)

7.44 R.R. Wills, J.A. Cunningham, J.M. Wimmer, R.W. Stewart: J. Am. Cer. Soc. **59**, 269-270 (1976)

7.45 R.N. Katz: Science **208**, 841-847 (1980)

7.46 J.K. Patel, D.P. Thompson: Br. Ceram. Trans. **87**, 70-73 (1988)

7.47 C.L. Quackenbush, J.T. Smith: Am. Cer. Soc. Bull. **59**, 532-537 (1980)

7.48 Y. Hasegawa, H. Tanaka, M. Tsutsumi, H. Suzuki: J. Cer. Soc. Jap. **88**, 292-297 (1980)

7.49 G.A. Weaver, J.W. Lucek: Am. Cer. Soc. Bull. **57**, 1131-1134, 1136 (1978)

7.50 C. O'Meara, J. Chen, J. Sjöberg, L. Pejryd: "Oxidation of Silicon Nitride", in *Structural Ceramics. Processing, Microstructure and Properties*, ed. by J.J. Bentzen et al. (Riso Nat. Lab., Roskilde 1990) pp. 439-444

7.51 J. Chen, J. Sjöberg, O. Lindqvist, C. O'Meara, L. Pejryd: J. Europ. Cer. Soc. **7**, 319-327 (1991)

7.52 J. Echeberria, F. Castro: "Comparison between Continuous and Cyclic Oxidation of Fully Dense Si_3N_4 + 1 w/o Y_2O_3", in *Structural Ceramics. Processing, Microstructure and Properties*, ed. by J.J. Bentzen et al. (Riso Nat. Lab., Roskilde 1990) pp. 249-255

7.53 N. Hirosaki, Y. Akimune, T. Ogasawara, A. Odaka: J. Mat. Sci. Lett. **10**, 753-755 (1991)

7.54 C. O'Meara, J. Sjöberg, G. Dunlop: J. Europ. Cer. Soc. **7**, 369-378 (1991)

7.55 J. Persson, M. Nygren: "Oxidation Studies of β-sialons", in *Structural Ceramics. Processing, Microstructure and Properties*, ed. by J.J. Bentzen et al. (Riso Nat. Lab., Roskilde 1990) pp. 451-456

7.56 J.A. Palm, C.D. Greshkovich: Am. Cer. Soc. Bull. **59**, 447-452 (1980)

7.57 J.A. Mangels: J. Mater. Sci. **15**, 2132-2135 (1980)

7.58 J. Desmaison: "High Temperature Oxidation of Nonoxide Structural Ceramics: Use of Advanced Protective Coatings", in *High Temperature Corrosion of Technical Ceramics*, ed. by R.J. Fordham (Elsevier, London 1990) pp. 93-108

Subject Index

CPSIA information can be obtained at www.ICGtesting.com
Printed in the USA
LVOW080313211112

308292LV00002B/14/P